Mathematics and Culture II

Michele Emmer

Mathematics and Culture II

Visual Perfection:
Mathematics and Creativity

 Springer

Editor
Michele Emmer
Dipartimento di Matematica "G. Castelnuovo"
Università degli Studi "La Sapienza", Roma, Italy
Piazzale Aldo Moro 2
00185 Roma, Italy
e-mail: emmer@mat.uniroma1.it

Library of Congress Control Number: 2003064905

A catalog record for this book is available from the Library of Congress.

Bibliographic information published by Die Deutsche Bibliothek
Die Deutsche Bibliothek lists this publication in the Deutsche Nationalbibliografie;
detailed bibliographic data is available in the Internet at <http://dnb.ddb.de>.

The articles by Apostolos Doxiadis *Euclid's Poetics: An examination of the similarity between narrative and proof,* Simon Singh *The Rise of Narrative Non-Fiction* and Robert Osserman *Mathematics Takes Center Stage* were originally published in Italian in Matematica e Cultura 2002, 88-470-0154-4 by Springer Italia, Milano 2002.

The Articles *Mathematics and Culture: Ever New Ideas; Mathematics, Art and Architecture; Mathland: From Topology to Virtual Architecture; Mathematics, Literature and Cinema* and *Mathematics and Raymond Queneau* by Michele Emmer were translated from the Italian Language by Gianfranco Marletta.
e-mail: gianfranco@marletta.co.uk

Mathematics Subject Classification (2000): 00Axx, 00B10, 01-XX

ISBN 3-540-21368-6 Springer Berlin Heidelberg New York

Springer is a part of Springer Science+Business Media
springeronline.com

© Springer-Verlag Berlin Heidelberg 2005
Printed in Germany

The use of general descriptive names, registered names, trademarks, etc. in this publication does not imply, even in the absence of a specific statement, that such names are exempt from the relevant protective laws and regulations and therefore free for general use.

Engraving on cover and part beginnings by Matteo Emmer, from the book: M. Emmer "La Venezia perfetta", Centro Internazionale della Grafica, Venezia, 2003; by kind permission.
Typeset: perform electronic publishing GmbH, Heidelberg
Cover design: Erich Kirchner, Heidelberg
Production: LE-TEX Jelonek, Schmidt & Vöckler GbR, Leipzig

Printed on acid-free paper   46/3142/YL - 5 4 3 2 1 0

Mathematics and Culture: Ever New Ideas

Michele Emmer

One of the interesting questions in the study of links between mathematics, art and creativity is whether a mathematician's creativity leads him to invent a new world, or rather makes him discover one that already exists in its own right. It could seem like a superfluous question of very little interest. Though it might seem so to non mathematicians, it certainly isn't so for many mathematicians like Roger Penrose, who has dedicated a part of the book *The Emperor's New Mind* [1] to this subject. "In mathematics, should one talk of invention or of discovery?" asks Roger Penrose. There are two possible answers to the question: when the mathematician obtains new results, he creates only some elaborate mental constructions which, although have no relation to physical reality whatsoever, nevertheless possess such power and elegance that they are able to make the researcher believe that these "mere mental constructions" have their very own reality. Alternatively, do mathematicians discover that "these mere mental constructions" are already there, a truth whose existence is completely independent of their workings out? Penrose is inclined toward the second point of view, even though he adds that the problem is not as simple as it seems. His opinion is that in mathematics one determines situations for which the term discovery is certainly more appropriate than the term invention. There are cases in which the results essentially derive from the structure itself, more than from the input of mathematicians. Penrose cites the example of complex numbers: "Later we find many other magical properties that these complex numbers possess, properties that we had no inkling about at first. These properties are just there. They were not put there by Cardano, nor by Bombelli, nor Wallis, nor Coates, nor Euler, nor Wessel, nor Gauss, despite the undoubted farsightedness of these, and other, great mathematicians; such magic was inherent in the very structure that they gradually uncovered."

When mathematicians discover a structure of this kind, it means that they stumbled upon that which Penrose calls "works of God". So, are mathematicians mere explorers? Fortunately not all mathematical structures are so strictly predetermined. There are cases in which "the results are obtained equally by merit of the structure and of the mathematicians' calculations"; in this case, Penrose says, it is more appropriate to use the word invention than the word discovery. Hence there is room for what he calls works of man, though he notes that the discoveries of structures that are works of God are of vastly greater importance than the 'mere' inventions that are the works of man.

V

One can make analogous distinctions in the arts and in engineering. "Great works of art are indeed 'closer to God' than are lesser ones."

Among artists, Penrose claims, the idea that their most important works reveal eternal truths is not uncommon, that they have "some kind of prior ethereal existence", while the lesser important works have a more personal character, and are arbitrary, mortal constructions. In mathematics, this need to believe in an immaterial and eternal existence, at least of the most profound mathematical concepts (the works of God), is felt even more strongly. Penrose observes that "there is a compelling uniqueness and universality in such mathematical ideas which seems to be of quite a different order from that which one could expect in the arts or or engineering." A work of art can be appreciated or brought into question in different epochs, but no-one can put in doubt a correct proof of a mathematical result.

Penrose explains very explicitly that mathematicians think of their discipline as a highly creative activity that has nothing to envy of the creativity of artists and indeed, because of the uniqueness and universality of mathematical creation, one that should be regarded as superior to the artistic discipline; Penrose doesn't write it explicitly, but many mathematicians think that mathematics is the true Art; a difficult, laborious art, with it's very own language and symbolism, that produces universally accepted results.

Penrose takes up the question of the existence in its own right of the world of mathematical ideas, a question that has dogged the science of mathematics since its inception. It comes from the book *Matière à pensée*, co-authored by a mathematician, Alain Connes, winner of the Fields medal, and a neurobiologist, Jean-Pierre Changeux [2].

In one of the chapters in the book, entitled *Invention ou découverte* the neurobiologist, speaking of the nature of mathematical objects, recalls that there is a realist attitude directly inspired by Plato, an attitude that can be summarised in the phrase: the world is populated by ideas that have a reality separate from physical reality. Supporting his observation, Changeux quotes a claim of Dieudonné, according to whom "mathematicians accept that mathematical objects possess a reality distinct from physical reality", a reality that can be compared to that which Plato accords to his Ideas. From this point of view it is of secondary importance whether or not the mathematical world is a divine creation, as the mathematician Cantor (1874–1918) believed: "The highest perfection of God is the possibility of creating an infinite set, and his immense bounty leads him to do so."

We are in complete divine mathesis, in complete metaphysics. This is surprising to serious scientists, comments Changeux. Connes, the mathematician, not at all bothered by the biologist's arguments, replies very clearly that he identifies strongly with the realist point of view. After emphasising that the sequence of prime numbers has a more stable reality than the material reality surrounding us, he notes that "one can compare the work of a mathematician to that of an explorer discovering the world." If practical experience leads us to discover pure and simple facts such as, for example, that the sequence of prime numbers appear to have no end, the work of a mathematician consists of proving that there exists an infinite number of primes. Once this property has been proved, no-one can ever claim to having found the largest prime of all. It would be easy to show them that they are wrong. Connes concludes "so we clash with a reality just as incontestable as the physical one". Indeed, it is arguably more real than any physical reality.

Still in the aforementioned book, we read that mathematics is a universal language, but that is not all, for the generative nature of mathematics has a crucial role. Otherwise one could not explain the incredible and unforeseeable usefulness of mathematics, from meteorology to the structure of DNA, to the simulation of boats for the America's Cup.

Imagination is not enough to do mathematics; you need knowledge of the language and techniques; imagination, creativity, the ability to make problems explicit, to formalise and solve them, whenever possible.

In recent years, thanks to the advent of computer graphics, the role of visualisation in some sectors of mathematics has increased significantly. It was inevitable that this sort of new *Visual Mathematics* would open new ways for links between creativity in the arts and in mathematics. We wish to deal with some of these aspects in this new book of the series called significantly "Mathematics and Culture", which was born of an idea of Valeria Marchiafava and Michele Emmer in September 1997, in Turin.

References

[1]   Penrose, R.: *The Emperor's New Mind.* Oxford University Press, New York (1989).
[2]   Changeux, Je.-P., Connes, A.: *Matière à Penseée.* ed. Odile Jacob, Paris (1989).

Table of Contents

IX

Mathematics, Literature and Cinema

Mathematics, Art and Architecture

Mathematics, Art and Architecture

Michele Emmer

The care that mathematicians have for the aesthetic quality of their discipline is noteworthy: this gives rise to many mathematicians' idea that mathematical activity and artistic activity are, in some sense, very similar and comparable.

Morris Kline, a mathematician, (and he is but one of many examples that we could cite) has dedicated several pages of his book *Mathematics In Western Culture* [1] to this subject. After recalling that mathematicians have for hundreds of years accepted what the Greeks maintained, that is, the fact that mathematics is an art and that mathematical work must satisfy aesthetic requirements, Kline asks the fundamental question of why many people maintain that the inclusion of mathematics in the arts is unjustified.

One of the most common objections is that mathematics does not evoke any emotion. Kline, on the other hand, observes that mathematics provokes undeniable feelings of aversion and reaction, and moreover generates great joy in the researchers when they manage to make a precise formulation of their ideas and on obtaining clever and inspired proofs. The problem lies in the fact that it is only researchers who can experience these emotions and no-one else. "Just like in the arts, each particular of the final work is not discovered, but composed. Of course the creative process must produce a work that has design, harmony and beauty. These qualities too are present in mathematical creations."

Commenting on these words of Kline, I have observed that [2] "if it is not interesting, in this realm, to discuss mathematicians' ideas on art, it is however worthwhile highlighting how this artistic aspiration is widespread in the mathematical community. Complementary to this requirement is the need for recognition of the artistic creativity of the mathematician by the mathematical laity; a recognition that is not generally given, in particular by those who take a professional interest in art. Especially since this would entail having to understand something of contemporary mathematics. Everyone can look at a work of art, listen to a symphony, but one cannot look at or listen to mathematics. Kline clearly recognises this when he says that the definitive test of a work of art is it's contribution to aesthetic pleasure or to beauty.

Fortunately, or unfortunately, the test in question is subjective, one that depends on the level of culture in a specific sector. The question as to whether or not mathematics possesses its own kind of beauty can thus be given a reply only by those who have a culture in this discipline ... Unfortunately, to master mathematical ideas requires study and there exists no direct route that substantially accomplishes this ".

References

[1]   Kline, M.: *Mathematics in Western Culture,* Oxford University Press, New York, 1953

[2]   Emmer, M.: *La perfezione visibile: arte matematica.* Theoria, Roma (1991)

From Tiling the Plane to Paving Town Square

Judith Flagg Moran, Kim Williams

Introduction

Every culture which has left us its artifacts has in so doing left evidence for the universal human fascination with patterns, and humans' propensity, if not compulsion, to ornament their environment, objects, and persons. [1] At the beginning of the current century, neuroscientists and cognitive neuropsychologists such as Stanislas Dehaene and Brian Butterworth are using formidable imaging techniques to actually watch the brain process numbers. But even a PET scan of a brain perceiving the pattern of a Peruvian blanket does not address the question of why this activity should be pleasurable or why the original weaver felt impelled to create the pattern in the first place. Writers on the psychology of art such as E. H. Gombrich and Rudolph Arnheim have addressed these questions by linking the perception of pattern to cognitive development through the perception of structure and so to human evolutionary success. In *New Essays on the Psychology of Art,* Arnheim states,

> *Perception must look for structure. In fact, perception is the discovery of structure. Structure tells us what the components of things are and by what sort of order they interact.* [2]

In A Sense of Order, Gombrich states his belief that the human sense of order is "rooted in man's biological heritage" and links pattern perception more directly to survival:

> *I believe that in the struggle for existence organisms developed a sense of order not because their environment was generally orderly but rather because perception requires a framework against which to plot deviations from regularity.*[3]

More recently, John Barrow in *The Artful Universe* echoes Gombrich in linking recognition of order in the environment with survival skills: recognition of order is beneficial for survival; survival requires being able to pick out the shape of the tiger against the pattern of the leaves in the jungle. At some point, however, the ability to recognize order in the environment, an adaptive skill, became an end in itself, that is, the recognition of order became pleasureable: aesthetics was born.

While accepting Barrow's basic idea that "Perhaps the most basic of all (human responses honed by natural selection) is an ability to sense and classify pattern" [4], we find two aspects of his argument particularly intriguing. First, that cognition as a biological trait is influenced over time, that is, it is "subject to evolution".[5] Second, that as our ability to perceive order has evolved, so has our sense of what constitutes order and pattern. This has been especially true in the latter part of the twentieth century, when we have come to accept new configurations as orderly because we have learned and accepted new criteria, in other words, we have extended and developed our "sense of order". Two of the most well-known of these new families of patterns are the fractals and the non-periodic patterns such as Penrose tilings and their descendants and relations. In part, generation and recognition of these patterns has been made possible through the revolution in information technology. Whereas previously, traditional patterns were generated by humans using the mechanism of repeated translation, the technique of rapid iteration, made feasible by the development of the computer, produces the resultant striking images of fractals and nonperiodic tilings in seconds. The instantaneous nature of modern communication technology disseminates the new images rapidly. And in the case of these two families of patterns, their use in modeling has ensured they were brought to the attention of a significant part of the general public. (Fractals can serve as the basis for the creation of natural features in computer graphics; Penrose tilings, discovered at the end of the 1970's, were used in the 1980's as a visual model for quasicrystals [6], the "new form of matter", the discovery of which was splashed across the pages of the *New York Times* and other media.) Once such patterns enter the domain of the general public, they become available to artists, who in turn bring them into the realm of aesthetics and make them part of their culture's pattern vocabulary. The role of artists in the integration of a sense of order into our environment is fundamental. One of the most pattern-obsessed of contemporary artists, M.C. Escher said in 1965:

> *Although I am even now still a layman in the area of mathematics, and although I lack theoretical knowledge, the mathematicians, and in particular the crystallographers, have had considerable influence on my work ... The laws of the phenomena around us – order, regularity, cyclical repetitions, and renewals – have assumed greater and greater importance for me. The awareness of their presence gives me peace and provides me with support. I try in my prints to testify that we live in a beautiful and orderly world, and not in a formless chaos, as it sometimes seems.[7]*

It is intriguing to imagine the use Escher (who died in 1972) might have made of the scaffolding for his work afforded by fractal and non-periodic patterns.

In this paper we have chosen to look at order through the vehicle of decorated pavements, that is, the application of geometric patterns to floors in architectural monuments and urban spaces. Decoration of buildings is common to all cultures and all time periods; the choice of paving designs often reflects cultural values about order. For instance, during the Renaissance, Leon Battista Alberti wrote,

And I would have the Composition of the Lines of the Pavement full of Musical and geometrical Proportions; to the Intent that which-soever Way we turn our Eyes, we may be sure to find Employment for our Minds. [8]

Alberti makes clear the Renaissance preference for the so-called harmonic proportions that were a hallmark of the architecture of that age. But it is significant that Alberti singles out the pavement for this prescription. The pavement often serves as a sort of canvas for the representation of ideas inherent in the architecture. After all, the floor is often the largest unbroken surface in a building. Paving patterns are not merely orderly in themselves, but contribute significantly to a larger sense of order in the built environment. They do this by indicating the organization that governs the building or space being decorated. Paving designs can indicate a hierarchy of spaces inside a building, they can indicate direction of movement through these spaces, and they can indicate a rhythm and speed of this movement. The present paper of course in no way serves as a survey of pavement design. Our intent is to contrast some pavement designs executed near the beginning of the second millennium with the pavements of two designers working at its end to illustrate how our sense of order and pattern has and continues to evolve.

The Pavements of the Cosmati

In spite of the wear and tear of almost a thousand years, Cosmatesque pavements still overwhelm the senses with their vibrancy, richness and variety – colorful carpets of marble that contrast with the austere simplicity of the Romanesque architecture that they adorn (Figure 1).[9]

The term 'Cosmatesque' refers to a particular style of polychrome decoration created through the use of tesserae or small tiles of marble, granite or ceramic to form geometric patterns. This kind of decoration takes its name from that of members of several families of artisans who created this type of ornament in the twelfth and thirteenth centuries in Italy. Cosmatesque pavements function as space organizers and direction indicators within their spaces. Romanesque churches are based on the basilican plan, that is, a longitudinal nave flanked by side aisles connects the entrance at one end of the church through the choir space to the altar placed before a round apse at the other end. Cosmatesque pavements are always composed of a linear element that marches up the nave and through the choir to arrive at the altar. This linear element may be composed of one or a combination of the two leitmotives of the Cosmati: the guilloche (Figure 2), a series of disks or roundels connecting by interweaving borders, and the quincunx, an arrangements of four roundels around a fifth, also connected by interweaving borders (Figure 3). Much of the rest of the floor area is subdivided into a grid of rectangles, each of which is filled with a geometric pattern that has a repeat in two directions, like wallpaper (Figure 4). (In fact, such patterns are commonly called wallpaper patterns, even by mathematicians.)

The pavement arrangement functions at two different levels. The linear pattern defines the space of the nave both as an architectural element, a corridor, and

Fig. 1.
The Pavements
of the Cosmati

8

Fig. 2.
Cosmati:
The guilloche

Fig. 3.
Cosmati: "The quincunx"

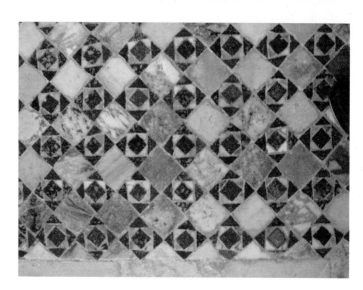

9

Fig. 4.
"Wallpaper pattern"

as a symbolic passage, emblematic of the earthly pilgrimage of the Christian before he enters heaven. The linear elements of guilloche and quincunx emphasize an uninterrupted progression. The sinuous bands linking the roundels are perceived as overlapping and continuous rather than disjoint and unconnected. In contrast to the patterns in the central nave, the geometric patterns within the rectangles that fill some or all of the remaining floor area are non-directional and static, providing a colorful, rich, marble carpet for the spaces.

The bilateral symmetry of the basilican plan is perhaps the most important ordering element of the architectural space. The Early Christian and medieval basilica was an interpretation based on Christian precepts of the exising architectural form of the Roman basilica. The Roman basilica was rectangular, with an apse on each end of the major axis and a doorways on each end of the minor axis. The architectural elements were arranged so that like elements are always opposite: apse to apse, column to column, doorway to doorway. In other words, the center of the Roman basilica was the center of rotation. To adapt the Roman basilica to their own ecclesiastical needs, Christian architects removed the entrances from the minor axis, eliminated the apse on one end of the major axis, replacing it with the entrance, and placed the altar in the remaining apse. Thus the symmetry of the plan was radically altered, as there remained only a single axis of reflection; the center of the structure no longer served as a center of rotational symmetry. This axis of mirror symmetry took on an all-important symbolic role: it became a path, as we said previously, symbolizing the Christian's pilgrimage to heaven. In Cosmatesque pavements, the curvilinear pattern that defines the axis of the nave emphasizes the mirror symmetry of the entire plan. This bilateral symmetry is further highlighted by the fact that in many cases the patterns in the rectangles on either side of central axis are arranged symmetrically, that is, the same pattern appears in rectangles in the same position on either side of the axis. Interestingly, besides being distinguished by their architectural roles, the central passageway and the rectangles adjoining it are also differentiated by the symmetries of their patterns. While the central element is itself an axis of mirror symmetry within the space of the church, due to the interwoven character of the guilloche and the quincunx neither exhibits mirror symmetry, but only rotational symmetry. That is, a turn of 90 degrees or one-quarter of a complete turn about the center point of the quincunx will cause it to coincide with itself. We say the quincunx has four-fold rotational symmetry. For a strip of three connected roundels forming a guilloche, a half-turn about the center point of the middle roundel will cause the entire strip to coincide with itself, and we say the guilloche has two-fold rotational symmetry. A pattern has mirror symmetry about a line if the line serves as an imaginary mirror reflecting one half of the pattern to generate the other. To see the distinction between mirror and rotational symmetries, consider the pavement design in Figure 5, from the side chapel of the Cardinal of Portugal in San Miniato al Monte in Florence. This is an interesting example of a pattern which looks superficially like a quincunx, but unlike a quincunx, it has four lines of mirror symmetry, and thus has the same symmetries as a square. The wallpaper patterns that fill the rectangles on either side of the aisle, on the other hand, unlike the designs in the central corridor, all exhibit mirror symmetry. As an example, the wallpaper pattern in Figure 4 has four different kinds of mirror lines: horizontal, vertical, and diagonal, oriented like those of the design in Figure 5. While not all the wallpaper patterns created by the Cosmati incorporate so many different kinds of mirrors, all have at least one set of parallel mirror lines.

Another interesting aspect of Cosmatesque pavements is the variety of shapes used in their realization. It is as though the Cosmati wanted to show their mastery of all the possible ways of tiling the plane that use mirror symmetry. The Cosmati took the marble for their pavements from the buildings of the Romans. The roundels used in the guilloches and quincunxes are slices from columns of antique buildings. Shapes appearing in the pavements include circles, triangles, squares,

Fig. 5.
Pavement design
in the side chapel
of the Cardinal
of Portugal in
San Miniato al Monte
in Florence

rectangles, rhombs, hexagons, octagons, and the *vesica piscis* (a pointed oval formed from two intersecting circles). Often one of these shapes is derived from another: a rhombus created from two equilateral triangles, a triangle generated from the division of a square along one of its diagonals, or a rectangle created by joining two squares. Cosmati patterns also tend to feature standard combinations of these forms, such as one square rotated 45 degrees and inscribed in another, (called *ad quadratum*), a triangle rotated 180 degrees and inscribed in another, (*ad triangulatum*), and a circle within another circle with the same center. As far as we have been able to determine, certain other combinations such as a square inscribed in a circle never appear in original pavements, (though this arrangement does appear in restored sections of pavements). We have chosen the term "vocabulary of shape" to characterize these particular tile shapes and the ways they were combined by the Cosmati. We use this term to emphasize that these craftsmen had rules for creating and combining these shapes in the same way that writers employ grammatical rules to determine what elements are used to create sentences. The elaboration of the Cosmati vocabulary of shape is the subject of a study on which we are currently working.

The geometric patterns of the Cosmati are often based in a very practical and constructive way upon the concept of space filling that is directly related to the method for laying the mosaic pieces to create the patterns. As shown by an incomplete fragment on display in the lower chapel of the church of Civita Castellana (Figure 6), the base piece of white marble was carved so that there was a raised border and a sunken tray to receive the pieces of colored marble that would form the wallpaper pattern. The sunken tray was filled with a cementitious setting bed, into which the tiles were pressed. The Cosmati artist began by setting the largest pieces of the pattern in an initial matrix, here a checkerboard pattern of squares. Thus, at the end of this first step, he has a checkerboard composed of solid square tiles and the spaces between them. His next step is to begin to fill the empty spaces with tiles of the next smaller scale, in this case a square whose diagonal is equal to the side of the larger square, in an *ad quadratum* pattern. At the end of this sec-

Fig. 6.
An incomplete fragment
on display in the lower
chapel of the church
of Civita Castellanae

ond step, he may choose to fill the remaining space with a piece that fits exactly, as the triangle used in this example, or, if the void left over is sufficiently large, he may choose to place a piece of the next smaller scale to partially fill that void before completing his spacefilling by inserting pieces that exactly fit the remaining spaces. This method of construction helps explain why, oddly enough for very colorful pavements, color is the least important element in the perception of order; indeed, there is no color symmetry in these pavements because color is often randomly applied. We read the patterns in terms of tone contrast, not color. For this reason it is possible to abstract the patterns as simple arrangements of black and while tiles, as art historian Dorothy Glass [10] did when classifying the patterns in order to date the pavements.

This space-filling imperative probably explains the frequent appearance of a third kind of symmetry in Cosmati patterns: similarity symmetry, or as it is often called, fractal symmetry. If leftover spaces are consistently filled with similar shapes of smaller scales, the result can be a pattern which is self-similar locally (that is, a section of the pattern possesses fractal symmetry.) Triangles are most frequently used as space-fillers, especially between the circular borders of a quincunx or a guilloche and the rectilinear border that surrounds it. This curved, almost-triangular space is often filled with a large triangle, and the remaining interstices filled with ever smaller triangles until no more filling is required. The resulting pattern is what would today be called a Sierpinski gasket (Figure 7).

Cosmati pavements thus exhibit a strong sense of order on several different scales: the overarching division of the church interior into passage and flanking rectangles, the rotational, mirror, and translational symmetries found within the rectangles and central element, and the self-similarity in certain areas which carries the eye ever deeper into the pattern. Color is used to dazzle and delight the eye, not to reinforce this symmetry, which we perceive mostly through tonal contrast.

Fig. 7.
A Sierpinski gasket

The Pavements of Carlo Scarpa

So what has changed in pavement design almost a thousand years after the Cosmati? We believe that the most significant development is a more subtle sense of what constitutes pattern, a more sophisticated "sense of order".

The pavement that Italian architect Carlo Scarpa designed for a relatively small space in the museum of the Palazzo Querini-Stampalia in Venice in 1961 exemplifies this development (Figure 8). [11] It is hard to imagine a pavement design more different from those of the Cosmati; indeed, those craftsmen would most likely have been puzzled and unimpressed with this contemporary floor. Even to modern eyes, a casual glance does not reveal an intended pattern on the part of the architect. Certainly there is no organization of the space, no implied direction

13

Fig. 8.
Pavement of Carlo Scarpa
in the museum
of the Palazzo Querini-
Stampalia in Venice,
1961

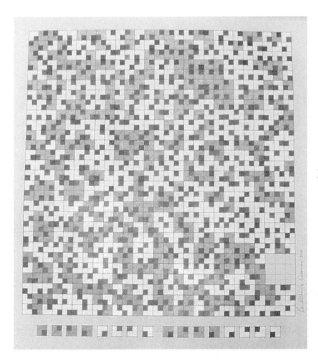

Fig. 9.
Pavement of Carlo Scarpa
in the museum
of the Palazzo
Querini- Stampalia

14

comparable to the central passage so characteristic of Cosmati pavements. A full appreciation of Scarpa's creation really requires a visit to Venice. Water is very much a part of the environment in the area where the pavement is located (when the tide is high, the water actually comes into the interior space), and the floor design mirrors and extends the motif and colors of the copper grating that separates the antechamber of the room from the canal. But apart from its superb functioning in its space, the floor exhibits a subtle design completely specified by Scarpa. The architect created the pavement using a pattern module or unit that he employed in at least three projects in the late twentieth century: the Palazzo Quernini-Stampalia, the church of the Torresino, Padua (1978) and the Castelvecchio Museum, Verona (1956). The module is a square, one quadrant of which is a smaller square of a contrasting color and/or texture; this can be read visually as an L-shape filled in with a small square. In the Palazzo Querini-Stampalia, Scarpa used two lighter marbles (cream and rosy beige) for the L-shape and two darker marbles (red and green) for the smaller squares, resulting in four differently colored modules. The four rotations of each of these four modules result in 16 differently oriented units. The overall pattern for the pavement is the result of the combination of these 16 units (Figure 9).

At first glance it would appear that the units in Scarpa's pavement are scattered randomly. However, his design drawings for the pavements show how carefully he arranged and color-coded the units so that nothing in the pavement's execution was left to chance. His drawings clearly indicate that Scarpa had thoughtfully arranged the units to produce an overall dappled effect. (This perhaps was in-

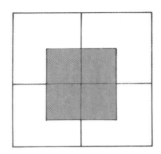 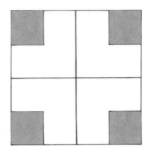

Fig. 10.
Symmetric
configurations

tended to reflect the dappled water of the canals just outside.) Even if we did not have the design drawings as a reference, however, one indication that the modules are not randomly arranged is provided by a consideration of the combinations that do *not* occur. For example, the first configuration in Figure 10 occurs only once, while the second is found only five times. Scarpa uses color to disguise these two highly symmetric groupings. In fact neither of the configurations in Figure 10 occur with all four L-shaped regions the same color. (We invite the reader to find other examples). If we distinguish configurations of four modules in a square only by the positions of the dark small squares, not taking color into account, there are 256 different possible configurations of four modules around a vertex. If the placement of the modules were random, we should then find in the pavement dozens of instances of each of the highly symmetric configurations shown in Figure 10. The fact that we see only one instance of the first pattern and five of the second, and these only when disguised by two different module colors, indicates that Scarpa specifically excluded them. It would seem that Scarpa sought to preclude any pattern arrangement that would attract or fix the spectator's eye on a given location. In *The Sense of Order* Gombrich claims "our eyes are attracted to points of maximal information content," such as the center of a design having rotational symmetry, or the mirror axis of a pattern. [12] Apparently Scarpa sought not to offer that "maximal information content" to the spectator.

Although Scarpa's pavement and those of the Cosmati are more different than alike, there is one symmetry that they have in common, though it appears to a greater degree in the Cosmati and imperfectly in the Scarpa – that of self-similarity. The L-shaped portion of Scarpa's module resembles the chair or L-tile, one of the rep-tiles of Solomon Golomb. [13] A rep-tile is a polygon which can be decomposed into congruent subtiles, each similar to the original. Figure 11 shows that both the L-tile and the equilateral triangle are rep-tiles. A rep-tile can be used to generate a tiling of the plane by decomposing the tile into its congruent subtiles, enlarging the resulting configuration until the subtiles reach the size of the original, decomposing each of the resulting tiles, and expanding again *ad infinitum*. (Of course, this procedure is better carried out by a computer than a twelfth-century artisan!) We see the first stages of this process in the Sierpinski gasket configurations in the Cosmati designs and, imperfectly, at isolated locations in the Scarpa pavement. Of course Scarpa did not use this method of decomposing and inflating to design his pavement for the Palazzo Stampalia-Querini. But because

15

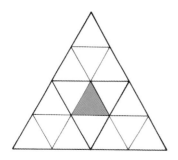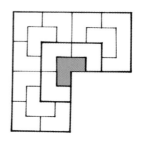

Fig. 11.
The L-tile and the
equilateral triangle
are rep-tiles

of the different colors of his basic square module, we can find many instances of a square region in the floor composed of four of his modules which contains an "L" of cream-colored squares, with an inset of a beige square, and even instances of a square area consisting of sixteen of his modules, twelve forming an "L" of cream squares, with a darker, beige square of 4 modules. This darker square, in turn, consists of an "L" of three beige modules and one cream module. Since the second and third-stage "L"s are decomposed into square, not L-shaped tiles, even these small regions are not examples of true rep-tiling.)

In addition to great differences between the high degree of symmetry in the Cosmati pavements and the more amorphous designs by Scarpa, the use of color also distinguishes the Scarpa pavement from those of the Cosmati. We have remarked that while color contrast in Cosmatesque pavements is very important, color itself appears to be randomly applied. On the other hand, in Scarpa's pavement, color is consciously used to create the basic repertory of sixteen distinguishable, differently oriented units.

Because we know that Scarpa's pavement was very carefully designed, it cannot be characterized as disordered, but it certainly reflects a different kind of order than that found in any Cosmati pattern. Perhaps the most intriguing aspect of Scarpa's pavement is its possible orderings that break down just when the eye seems to have grasped an underlying rule.

If Carlo Scarpa's pavement design in the Palazzo Querini-Stampalia refuses description in terms of conventional symmetries, if it can be said to be *almost* orderly (that is, some properties are found in it at some places, but not always in a regular fashion), can it still be correctly classified as pattern? Perhaps the answer lies in what Gombrich calls "the most basic fact of aesthetic experience", that "delight lies somewhere between boredom and confusion."[14]

The Pavements of Tess Jaray

Tess Jaray is a contemporary British abstract artist who has long worked with pattern and rhythm in her oil paintings and prints. In the late 1980's she turned her attention to pavement designs for several major metropolitan redevelopment projects in Great Britain, producing great urban works such as the Wakefield Cathedral Precinct and Centenary Square, Birmingham. Jaray's pavements have a much more classic feel and look to them than those of Carlo Scarpa; the spectator would never for a moment think the placement of bricks in one of her designs was random. Whereas the primary function of the Scarpa pavement examined here is to provide an identity for a single space, Jaray's pavements function much as did those of the Cosmati, identifying significant spaces, establishing hierarchies of space and indicating foci for the spectator's attention. As opposed to the rich variety of shapes used by the Cosmati, Jaray truly uses a minimalist's vocabulary of shape, that is, she uses one figure only, a simple rectangular brick with proportions 1:2. She tells how her patterns arose from the brick unit itself:

> The most common brick (although of course they vary) is used in the proportion 3:1, i.e., three bricks on their side can be fitted into the length of one. … In order to achieve ornamental patterning with this proportion, a high degree of skill was needed to infer any overall movement that was not only vertical or horizontal, but implied a diagonal that gave the surface a sense of dynamic … With the new production of the brick pavers, however, a new visual dynamic was made possible, at least for paving on the horizontal. In order to provide a brick with stronger structural properties that would allow for vehicular as well as pedestrian traffic, they were produced with the proportion 2:1. This allows for a very different geometry to be brought into play. [15]

In spite of the simplicity of Jaray's vocabulary of shape, she is able to achieve rich designs.

For the Wakefield Cathedral Precinct (Figure 12), Jaray designed a pavement that would help integrate the Gothic Cathedral with the commercial urban context that has grown up around it. She did this by designing two different patterns to define and identify the commercial and the cathedral zones, while at the same

Fig. 12.
Wakefield Cathedral
Precinct

Fig. 13.
Straightforward herringbone
brick pattern

time knitting both spaces together through the common scale of the bricks and the repetition of colors to provide unity for the entire urban setting. She distinguishes the cathedral plaza through the use of a pattern based on crosses, while the commerical shopping street features a geometric pattern with no symbolic content. Because of the scale of these large urban spaces, the patterns themselves must be very large. One simple way to increase the size of the perceived pattern is to introduce color contrast to a simple overall pattern of bricks. In the pattern for the shopping street in Wakefield, colored bricks enliven a straightforward herringbone brick pattern to provide visual variety and an increased awareness of pattern (Figure 13). In the plaza immediately adjacent to the cathedral itself, the pavement features a cross pattern. Underlying the apparently simple cross pattern is a more complex arrangement of bricks. The tiling pattern of the bricks themselves is somewhat obscured because the color contrast between the buff colored bricks that form the crosses and the deep indigo blue of the Harlech bricks of the background is so strong, but the brick pattern nevertheless adds an important element of scale by means of which the spectator is able to visually measure the urban space of the plaza.

Figure 14 shows just how important this contrast is to reading the pavement design. Figure 14a shows the pattern of bricks that tiles the cathedral precinct; Figure 14b shows how Jaray's use of colored bricks enables our eyes to pick out the underlying cross motif.

Jaray has written about the importance symmetry has for her:

> *Symmetry has played a part in this. It has been a fundamental element in the work since the beginning. It is present in various ways – direct and obvious, alluded to obliquely, implicit rather than explicit, so that when one symmetry is revealed, others should be concealed.* [16]

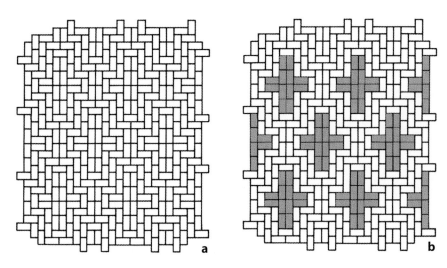

Fig. 14 a, b. Pavement designs

In this case the underlying pattern of bricks is concealed by the more obvious pattern of light crosses on a dark field. But the tiling of bricks itself, divorced of any color or contrast, is fascinating. A mathematician uses the word *tiling* to characterize a covering of a surface (in our discussion, a plane) by closed shapes without gaps or overlaps. This is also, of course, just what is desired by anyone paving a floor or plaza. Scarpa's basic module is a square; tiling the plane or paving a floor with square tiles is surely the easiest solution to filling the space. Scarpa's art is found in his use of color and the darker subsquare "decoration" on his module to create his overall design.

Jaray's tiling, like Scarpa's, is *monohedral;* that is, all her tiles are the same size and shape; they are all bricks in the proportion 2:1. But with this basic tile shape she creates a wealth of different tilings of the plane. The tiling in Figure 14a has the same symmetries as the perceived pattern of buff crosses on a blue field. Both have vertical and horizontal mirror lines or axes that intersect in the centers of the crosses and mid way between their horizontal arms. These intersecting mirror lines create centers of 2-fold rotation. The tiling contains two more kinds of 2-fold rotation centers. These occur at the mid-points of the two line segments connecting the center of a cross to the centers of the two crosses diagonally above it. These last two rotation centers do not lie on mirror lines; they are in fact, mirror images of each other.

In her pavement design for Centenary Square Birmingham, Jaray incorporates both all-over wallpaper designs and an overarching organization of this vast space about a strong central element (Figure 15). The Cosmati's central element led to the church altar; Jaray's leads the eye to the green space at the end of the plaza. The Cosmati central element was flanked by rectangular patches, each with a different wallpaper pattern; Jaray applies several wallpaper patterns to successive collars of a strong central corridor. The mirror axis of her pavement design is

19

Fig. 15. Pavement design for Centenary Square Birmingham

reinforced by the pattern in the central core: the centers of the row of motifs along the mirror axis of the pavement are colored differently from the rest of the motifs in this central area. Thus, unlike the Cosmati pavements, the central element in this pavement has mirror symmetry. But perhaps the most striking difference between this pavement and a Cosmati church floor is that all of Jaray's patterns are created by introducing color into a wealth of distinct tilings, all by bricks of the same 2:1 proportion. Within this restricted vocabulary of shape, Jaray is a virtuoso at creating different and interesting tilings. She has managed to work creatively within the constraints of the standardization of paving materials, and urges that other artists be enabled to do likewise:

> *Artists' understanding of the abstract visual language inherent in the geometry of brick building has been insuffiently used. Bring them in to invent ways of re-humanizing the use of materials when production becomes mechanised and standardised. But there are times when standardisation can actually encourage new direction ... This requires more thought and more atttention but can at least avoid those prosaic acres of herringbone which blight some of our urban centres. A dynamic or harmony beneath our feet will also create a heightened awareness of our surroundings.* [17]

Jaray's pavements are an illustration of how decoration adds to our sense of order in the environment, that is, the power of ornament when imaginatively applied to add the necessary structure for reading the world around us.

Conclusion

Today we continue to share with our forebears a delight in pattern. But pattern in architectural ornament is not only beautiful, it is useful as well. According to architectural theorist Christopher Alexander, "The main purpose of ornament in the environment – in buildings, rooms, and public spaces – is to make the world more whole by knitting it together"[18]; "When (ornament) is well used, it is always applied in a place where there is a genuine gap, a need for a little more structure".[19] The pavements discussed here admirably serve the purpose of knitting their own worlds together: the Cosmati knit together the hierarchical spaces of the Christian basilica; Carlo Scarpa knits together the interior of a Venetian museum with the dappled waters of the canals just outside; Tess Jaray knits together very large scale urban spaces that without pattern might be only sterile and intimidating as so many twentieth century spaces are.

But we differ from our forebears in that our ability to recognize and design patterns has evolved through the centuries and what we today recognize as pattern (fractals, for example) would have not been perceived as pattern in the past. As John Barrow writes, "Only with the coming of studies of the complex, by means of new technologies, has the scientific eye turned to the problem of explaining diversity, asymmetry, and irregularity", pointing the way towards the development of ever more sophisticated patterning and order. [20]

Acknowledgment

Moran's research for this paper was partly funded by a grant from the Graham Foundation for Advanced Studies in the Fine Arts, Chicago.

References

[1] See D.W. Crowe and D.K. Washburn, *Symmetries of Culture: Theory and Practice of Plane Pattern Analysis* (Seattle, WA: University of Washington Press, 1988).
[2] R. Arnheim, *New Essays on the Psychology of Art* (Berkeley: University of California Press, 1986) p. 253.
[3] E.H. Gombrich, *The Sense of Order: A study in the psychology of decorative art* (Ithica NY: Cornell University Press, 1984) p. xii.
[4] J.D. Barrow, *The Artful Universe* (Oxford: Clarendon Press, 1995) p. 104.
[5] Barrow [4] p. 28.
[6] See M. Senechal, *Quasicrystals and Geometry* (Cambridge: Cambridge University Press, 1995).

21

[7] M.C. Escher, *Escher on Escher: Exploring the Infinite* (New York: Harry N. Abrams, 1989) p. 21.

[8] L.B. Alberti, *The Ten Books of Architecture*. 1755 (reprint, New York: Dover Publications, 1986) p. 150.

[9] For more on the Cosmatesque pavements, see K. Williams, *Italian Pavements. Patterns in Space* (Houston: Anchorage Press, 1997) and K. Williams, The Pavements of the Cosmati. Pp. 41–45 in *The Mathematical Intelligencer 19* (Winter 1997): 1, 41–45.

[10] See D. Glass, *Studies on Cosmatesque Pavements,* British Archaeological Reports International Series, Vol. 82 (Oxford: British Archaeological Reports, 1980).

[11] For Carlo Scarpa's designs for the Querini-Stampalia, see F. Dal Co and G.Mazzariol, *Carlo Scarpa 1906–1978* (Milan: Electa, 1984); S. Los and K. Frahm., *Carlo Scarpa* (Cologne: Benedikt Taschen Verlag, 1994); M. Mazza, Marta (ed.), *Carlo Scarpa alla Querini Stampalia. Disegni Inediti* (Venice: Il Cardo, 1996).

[12] Gombrich [3] p. 121.

[13] B. Grünbaum and G.C. Shephard, *Tilings and Patterns* (New York: Freeman, 1987) section 10.1. See also D. Schattschneider, "The Fascination of Tiling", in M. Emmer (ed.), *The Visual Mind* (Cambridge MA: MIT Press, 1993) pp. 157–164.

[14] Gombrich [3] p. 9.

[15] K. Williams, "Environmental Patterns: Paving Designs by Tess Jaray", in *Nexus Network Journal* 2 (2000): 87–92, pp. 88–89.

[16] T. Jaray, personal correspondence.

[17] T. Jaray, "The Expressive Power of Brickwork", in *Architects Journal* 6 (November 1997) pp. 6–7.

[18] C. Alexander, 1977. *A Pattern Language* (New York: Oxford University Press, 1977) p. 1149.

[19] Alexander [18] p. 1151.

[20] Barrow [4], p. 245.

Mathematics in Contemporary Arts – Finite and Infinity

Dietmar Guderian

Introduction

Two thousand years ago a paper with the topic "mathematics and art" would not have been possible in the form of today, because mathematics and arts together were muses.

Just in modern times arts and mathematics became different disciplines with different profiles of thinking, working methods and aims.

Considering world-wide networks, the unbelievable fullness of data and facts and the necessity to structure our environment intellectualy, it seems to be useful to track down similarities, common strategies etc. in the different diciplines today.

Mathematics and art – here the descriptive art – represent only one from various different couple formations.

The article will point out only one aspect of all the artworks which will be shown.

There are lots of ways to a piece of art. The way which uses links to mathematics is only one of them. All the other important aspects will be ignored here.

But this way differs from many others because it is characterized by typical characteristics of mathematics like explicability, explainability, understandability and inspectability.

That's exactly the attraction for a mathematician to deal with the subject mathematics and culture, in that case with mathematics and arts.

The article will be structured in the following way: at first short definitions of both disciplines will be introduced. The second part will demonstrate some typical historic and contemporary connections between mathematics and arts. And at the end, by way of example shall be demonstrated with the term Inifinity how different mathematics appear in contemporary art.

Mathematics and Arts

In dictionaries and other compendia (Rutherford) mathematics is defined as follows:

Mathematics studies structures and relationsships between objects of our thinking (numbers, variables a.s.o.) and patterns.

What is Art?

Art is the name of the totality of all things which are produced by men and which are not unique and totally defined by a function.

Like mathematics art doesn't invalidate preceded works. And it is known that esthetic aspects play an important role in searching intuitively for a proof. For good reason mathematicians often speak about elegancy, too (Dreyfus a.o.).

Still two characteristics absolutely don't apply to mathematics:

- All that is or was respected as art is dependent from the standards that are valid in one epoch.
- Art has not to proof its correctness.

Links Between Mathematics and Arts

Historical and Contemporary Links

The connection between mathematics and arts is not new at all. In the past mathematics was often used as a tool to create artworks: The Pythagorean school need a good ratio between the main lengths of sculptures, houses or temples to have a guarantee for harmony in them. Fibonacci-series, the Golden section and the development of the perspective mapping are examples for that, too. In our days elements of mathematics themselves may become contents of a piece of art – for instance the square of Malewitch.

At first we point out for some different fields of mathematics how they enter into arts.

Logics – especially negotiation and multiple negotiations show pictures of Vasarély in the thirties but today we find them at young artist, too, for instance in the work of the German artist Uwe Kubiak.

Congruent mappings are to be found in all early cultures in connection with their religions as axial-symmetry (christian cross), enlargement and reduction (pictures of pharaohs, portraits of sponsors in medieval Churches), rotation (windows of gothic and romanic churches) and translation (warriers in tumbs of pharaohs, arcades in romanic churches)

The use of plane and spatial forms does not in any case assume the use of numbers, too: magic circles, the triangle as a symbol for the holy ghost, the classic Islamic mosque with its construction out of a cube and half a sphere, the cubes of the romanic basilica, Egypt's pyramids.

But also elements of topology (labyrinths, ornaments, perquets) often do not presume any arithmetic facilities of the artist. Even medieval perspective mappings, Golden sections of the renaissance and modular buildings of statues and temples in classic Greece where possible without the use of calculations.

Until today artists realise artworks in this way: Dan Flavin in his axial-symmetric neon-installation at the Hamburger Bahnhof in Berlin, Claes Oldenburg with his oversized "documenta"- tool in Kassel, Bruce Naumanns video-portrait which rotated during the DOCUMENTA IX in Kassel. During the Biennale 1995 in Venice the corean artist Soocheon Jheon exposed series of identical human figures.

Three-dimensional geometry gives a lot of contemporary artists a wide working place, too.

The Swiss Max Bill shows all different halves of spheres at the university of Karlsruhe (Germany) and all different halves of cubes at Haifa (Israel), the German artist Manfred Mohr experimentates with hyper-cubes; the documenta-artist Isa Genzken builds ellipsoids etc..

Topology plays a role in the art of today, too: M.C. Escher shows the endless Moebius band.

The German Eberhard Fiebig and the Spanish Eduardo Chillida work with knots, open and closed curves, nets and other contents of the topology.

Pablo Picasso and Georges Braque began to open the wide field of changing perspective mapping.

Common to all of these artworks is: mathematical methods give them order.

Concrete Art

In Zürich (Switzerland) in the late thirties of the last century a group of artists (Max Bill, Camille Graeser, Verena Loewensberg, Richard Paul Lohse, …) started to search for ways to visualize abstract contents as for instance loneliness, order, disorder … They realized this by using only objective structures and relations as far as possible and they did'nt allow themselves to introduce any subjective elements into their works. Within some years the contents became more and more abstract. And in the end often the titles of their artworks were completely mathematical ones. Here as an example we discuss a piece of Anton Stankowski, who was one of the members of the famous group of the concrete artists of Zürich, too.

The artwork of Anton Stankowski (1906–1998) gives a good example to show how members of the first generation of concrete artists were able to introduce elementary geometrical contents into their pieces to give them order:
– Each side of the square is divided into four equal parts. The middle of each side is connected with one corner by a straight line. Always two of these lines form a

25

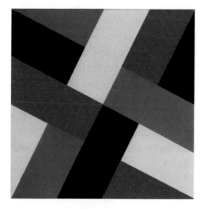

Fig. 1. Anton Stankowski

Fig. 2. Rotational Symmetry

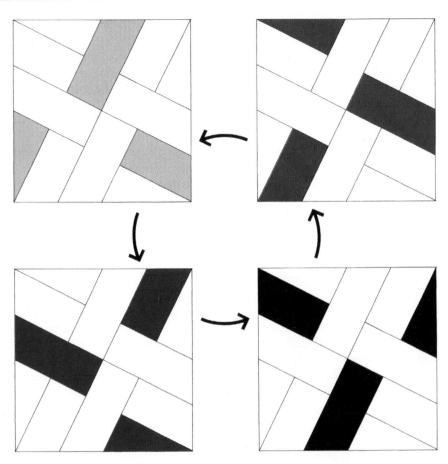

Fig. 3. Cyclical changing of colours

stripe. These stripes again are divided into two halves. Parts of all lines are hidden in such a way, that the picture gets a rotational symmetry with 90 degrees (Figure 2)

- It is easy to find out, that each of the three different shapes appears four times in the picture (Figure 2). By this it is possible to colour three times each of the four congruent shapes in one of four possible colours.
- Anton Stankowski places the four colours thus, that they change their places in a cyclic way when the picture is turned by 90 degrees (Figure 3).
- All shapes of the same colour are bount together in a special net: Following their border lines one finds: Each pair out of three shapes with same colour have one common border line together (Figure 4)
- Even the distribution of the colours during a "walk" around the picture is fixed strongly: They follow each other in a cyclic way, too: blue – black – red – yellow – blue- black ... (Figure 5).

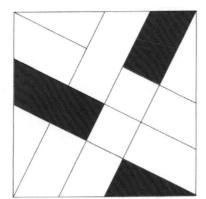

Fig. 4. Relations

At the first view the uninformed spectator might believe that the artist did not need much reflections before he made this artwork. But in fact it is the only possible piece: If one uses the same partition lines which the artist used and wants to place the colours in his way with equal area and with the possibility of a cyclical changing of colours one will find sixteen different pictures. But only the one which the artist worked out has all the additional properties which here were pointed out: All figures of the same colour are bcount into a net of common border lines (see above), and the colours follow each other in a cyclic way if a spectator surrounds the picture. The sixteen possible artworks are shown in Figure 6.

27

Fig. 5. Cyclical series

Fig. 6 a.
Eight of
Sixteen
possible
artworks

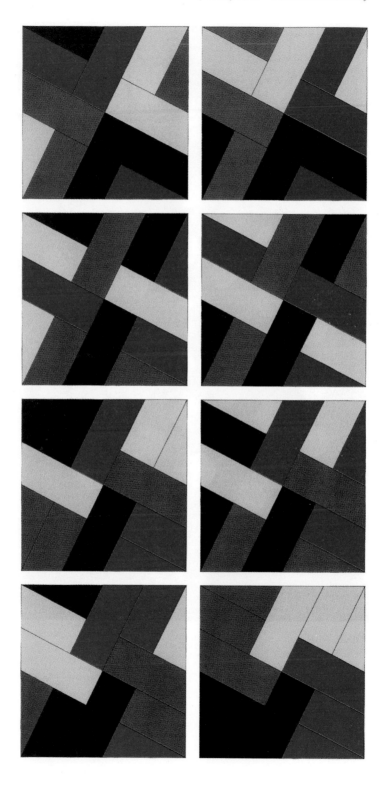

Fig. 6 b.
Eight of
Sixteen
possible
artworks

Fig. 7. Caris

Examples for Parallel Developments in Mathematics and Arts

It is exciting to find parallel developments between mathematics and arts. Like for example in the beginning of the chaos-theory: The artist Vantongerloo announces in the thirties the "heretical" opinion: "I have two solutions of one problem". At that time for the classical mathematics was valid: A problem is solved whenever you have proved the existence of a solution and its uniqueness. It took some more years until Goertler (university of Goettingen) and American Taylor laid the base of the chaos theory, when they showed simultaniously by experements first, that it was possible that the system of the Navier-Stokes-equations under special circumstances could have more than one solution. It is fascinating that an artist reaches in occupation with his art the same central sentence of a new scientific theory than the mathematicians and other scientists in their field.

Some decades later we experience a similar work side by side, as Salvador Dali paints pictures that open unexpected different changing contents of the same painting; for instance the painting of a dog „tips over" to the face of a girl which „tips over" to a copy of a goblet. As Dali heard that the mathematician Thom works out a mathematics theory of irreversible processes – so called catastrophes – he invited him. During long discussions they worked hard to find out the common structures of the results they had found in their different areas. At the end Dali adopts fascinated the mathematical functions and curves in his work. Finally Dali's last work shows the "tail of a swallow-curve" of one of Thoms elementary catastrophes.

The Dutch artist Gerard Caris gives another example for such parallelism (Figure 7): About thirty years ago he declared that the human society determines itself in all its thinking and doing because it favorises the rectangle too much. His idea was, that the use of the pentagon especially with respect of the manifold included Golden sections would be much more sensful. Though we all knew, that it is impossible to pattern the plane with the regular pentagon, he tried it again and again. He imported little regions of incorrectness into his works and at the end (in the early seventies) he created lots of artworks with local pentagonal symmetries. He tried the same with the dodecahedron and filled the space with them nearly completely though we knew that the packing of the space by dodecahedrons is theoretically impossible too. But Caris found local packings. It took about ten years more until scientists especially chemists found crystalline structures which included local five-symmetries, too. It was the birth of the theory and later the production of quasi-crystals.

Artworks Between Finite and Infinity

Only a few examples shall show how artists include today mathematics in their works and how different their ways may be even if they try to visualize nearly the same abstract contents. With view on the millennium we focus on some aspects of the abstract content "INFINITY" in the arts of nowadays.

The last piece of one of the famous founders of concrete art, Max Bill, was a Moebius band ("Endless Ribbon", red assuan granite, 235 cm × 190 cm × 130 cm,

1935–1996) which was installed in 2000 at the new museum in Vaduz/Liechten-stein: Max Bill for himself found the Moebius band during his work as an artist and he was astonished to hear, that Moebius found it earlier. But later on Bill made many sculptures with the Moebius band, which fascinated him because it had only one surface and one edge. Under the aspect of the infinity here it is to be noticed, that endless ways on a determined sculpture are possible.

With aide of combinatorics artists often try to find a big number of possibilities to create artworks. But as soon as the set of elements which shall be combined is finite, the number of different possibilities to combine them is finite, too. The American artist Sol Le Witt for instance superposes a spatial grid with a finite number of grid points over a cube. By methods of combinatorics he chooses a set of points of them and the shapes between them form an incomplete cube. There are a lot of different sculptures possible, but nevertheless their number is finite.

The German artist Bernhard Sandfort superposed four orthogonal nets of four different colours in his artwork. This artwork consists of sixteen little squares,

Fig. 8. Sandfort

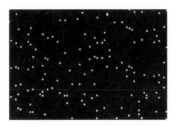

Fig. 9.
Sieve of Eratosthenes

and the artist in fact had a unbelievable great number of possibilities to put these sixteen little squares into a 4 × 4-square. But he accepted only three of them as artworks. He need a special relation between order and disorder in his pictures. For him this order existed as soon as two of the four nets are not disturbed. Figure 8 shows the artwork where the orange and the brown nets are not disturbed.

Rune Mields is a well known German female artist who ignores the existence of hazard. She is of the opinion that people especially scientists like to speak of hazard whenever they have not enough information about the things which stand behind their observations. To let the spectator find this out himself she created pictures which at the first view seem to show objects which are distributed by chance, but where she gives the solution afterwards, which shows everybody that there exists a strongly determined rule to place every object in the picture. As an example we show the "Sieb des Eratosthenes" (sieve of Eratosthenes) here (Figure 9 and Figure 10). If one looks first only to the part of the artwork which is shown in Fig-

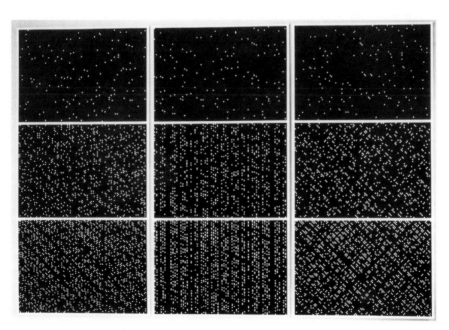

Fig. 10. Sieve of Eratosthenes

ure 9, then the white points in the picture seem to be distributed by chance. And nobody could find out with any method of statistical tests that these points are not placed by chance. But in fact here we see the distribution of prime numbers within a series of about 9000 integers in the range of 10(exp. 30). In the lowest row of the complete 9-part-artwork there are the prime numbers in the first 9000 integers marked (with 99, 100 and 101 integers in one row from the left to the right). We see the well know distribution and structure in this range. One step higher we see the distribution of prime numbers in a series of the range of 10 (exp. 6). The uppermost three parts of the installation do not show a structure any more, so that spectators - mathematicians, too – often speak of a distribution by chance as long as they do not know the background.

The Austrian artist Waltraut Cooper transforms the alphabet into the ASCII-Code for a long period of her working. For instance the word "Licht" (light) she transformed into a 5×5-Matrix of green and blue lightning sticks which formed a fleet during the Biennale of Venice in 1995. During the annual meeting of the German Society of Mathematicians in 1994 while the winner of the Nobel-price Eigen gave a lecture about information in the Mercator-hall at Duisburg, Waltraut Cooper showed the words of this lecture written in the ASCII-code realised by blue and green neon lights. For passengers in the street the green-blue changing lights on a 30m-wall of the hall seemed to be switched by chance. Only those people who were informed, knew, that the colours had to follow to a strong partiture.

Only a few artists found a way to introduce the infinity of the process of building a fractal and the self-similarity of fractals into artworks. The Swiss artist Karl Gerstner is one of the most important of them. His way to bring self-similarity into his artworks is the following one: He inscribes into a given circle the two biggest possible circles of equal area. One of them is filled by the same procedure as above and so on. This should be done up to the infinity. But for a painter it is impossible to paint infinitely small circles. Gerstner – a famous expert of the theory of colours – found a special way to give the imagination of an infinite process to the spectator: From one generation of circles to the next smaller one the difference between the colours of them become smaller and smaller. At the end the colours are so close to each other that it is possible to imagine that the division of circles goes further to infinity ("Color Fractal 8 (IA), Hommage B Mandelbrot", 1976/ 1988, collection Ruppert, Berlin).

Roman Opalka, a Polish artist who represented Poland during one of the Biennales at Venice, found a way to demonstrate, that time is infinite but the time for a human being on earth is finite: He started in 1965 to note the running of the time by writing the integers. He started with the number 1 and works nearly every night. In the morning he always takes a photo of himself. After about forty years he is in the region of six millions now, a number unbelievable small in face of the infinity of the number of integers. But the photos the artist took of himself in the years 1965 till 2000 show the dramatic difference between the infinity of the number of integers and the finite of our life.

35

Fig. 11. Cooper

Fig. 12. Ciervo

Costantino Ciervo is an Italian artist whose works were shown in the year 2001 for the first time in the USA (New York). He shows us that in his opinion there are lots of possibilities for the "truth": March 2000 during the annual meeting of the German Society of Didactics of Mathematics he installed a twelve meter long piece on a wall of the Altes Rathaus Potsdam. Sixty-four boxes were connected each of them by six wires to a symbolic truth-giving machine. Each wire could bear electricity or not – therefore 2(exp. 6) = 64 different "truths" were possible. But in addition to these 64 possibilities which were realised by combination of the different inputs he introduced a very special effect: Directed by photo-cells each box could change its worth of "truth" or not in the moment when a spectator passed. By this the artist demonstrated that in his opinion there does never exist only one objective "truth". Truth is always an individual coloured one. Reflecting the number of living individuals on earth we come to a very big but again not infinite number of "truths", too.

Summarisation

With only a few of examples we demonstrated how useful mathematics and arts are cooperating today. But links do not only exist between arts and mathematics. We find them between mathematics and music, mathematics and poetry and others, too. To study and use these links may give us tools to understand future connections between our science and the culture of tomorrow.

Literature

Max Bill, *Max Bill - Endless Ribbon 1935–1995 and the single-sided surfaces* (with texts from Max Bill, Jakob Bill, Dietmar Guderian and Michael Hilti, Benteli Verlag, Wabern-Bern, 2000.

Costantino Ciervo, *catalogue,* Kunstverein Hürth, 1999.

Costantino Ciervo, *catalogue,* Janus Gat Gallery, N.Y. in cooperation with Galerie Ralf Vostell, Berlin, 2000.

Waltraut Cooper, *Eine Werkübersicht,* Galerie im Stifterhaus, Landesgalerie Oberösterreich, Verlag der Apfel, 1998.

T. Dreyfus, Th. Eisenberg, *Symmetry in Mathematics Learning,* Zentralblatt für Didaktik der Mathematik (1990) vol. 2, p. 53.

S. Fehlemann, V. Flusser, K. Gerstner, *Karl Gerstner, catalogue,* Museum für Gegenwartskunst Basel, 1992.

K. Gerstner, *Ideenskizzen und Bilder – Color Fractals,* Edition Cantz, Ostfildern, 1992.

D. Guderian, *Über die Mathematik im Werk von Rune Mields,* Staatsgalerie Baden-Baden und Kunstverein Bonn, 1988.

D. Guderian, *Logik und Strenge in Anton Stankowskis Werk,* Johann-Karl Schmidt (ed.): Anton Stankowski: Gemälde 1927–1991, Galerie der Stadt Stuttgart, 1991.

D. Guderian, *Mathematik in der Kunst der letzten dreißig Jahre,* edition und galerie lahumière, Paris und Bannstein-Verlag, 79285 Ebringen i. Br. (English translation: *Mathematics in the Arts of the recent thirty Years,* see under www.ph-freiburg.de/mathe/guderian).

D. Guderian, *Mit System zu Wohlgefallen – Beziehungsgeflechte,* Herbert W. Kapitzki and Fritz Seitz (ed.): Anton Stankowski, *Freunde erinnern sich,* avedition Ludwigsburg, 1999.

D. Guderian, *Endlich – Unendlich* (catalogue),University Potsdam, Bannstein Verlag, 79285 Ebringen i. Br., 2000.

D. Guderian, *Der Erforscher Gerard Caris,* Gerard Caris (cat., ed. Peter Volkwein), Museum für Konkrete Kunst Ingolstadt, 2000.

F. James Rutherford (dir.), *Benchmarks for Science Literacy,* Oxford University Press, New York, Oxford, 1993.

B. Sandfort, *catalogue,* Städt. Kunsthalle Mannheim, 1987

De Insana Geometria

ACHILLE PERILLI

In re-examining those visual experiences which, during the course of the century, have been generically labelled as abstractionism, one seems to perceive, amongst the few theoretical certainties and the many poetic uncertainties, a thread linking the attainment of rationalist truth to the decadence of the same and its formal dissolution.

The slow but constant shift from representation to presentation, based on the *linguistic* values of the code and the denial of iconographical communication, undoubtedly freed the artist from various influences which bound him to the idea of reality. The visual factor acquired a degree of autonomy with respect to the concept of reproduction and was no longer intended as evidence, a record or window on history, but as a manner of investigating the nature of the world and of mankind's existence. Thus it becomes an analysis of the structure of the visual world; as Klee wrote, "Art does not copy visible objects, but renders them visible".

This progress and occasional regression, this lengthy voyage was meant to illuminate, uncover and reveal the most subtle workings of vision; to reproduce the procedure of focusing the image; to communicate it by means of an analytical methodology; and, above all, to extract from the depth of our memory, the richest and most complex forms of knowledge. This process knew both absolute certainties and shattering failures, due to the difficulties of carrying out an enquiry comprehensive of the many possible interpretations available to investigators. One of the absolute certainties was the belief that to proceed on the basis of pure perception (the retina) was enough to rediscover the essential laws of vision; all other approaches were dismissed as spurious or insufficiently pure. This was a certainty which derived from the conviction that everything could be reduced to the essence of painting, which coincided with the essence of geometrical form – the fundamental crossing of the horizon with a vertical line on a white surface.

To reach this point it was necessary to reduce painting, or the investigation of painting, to an absolute. Beyond this absolute, the value of representation was no longer intended as a transcription of reality, or continued by the necessity of transmitting a specific message (landscape, still life, portrait): it became a free component of pictorial activity, strengthened by the multiplication of *languages* through the ambiguity of speech (De Chirico, for instance, or Pop Art).

The equally drastic alternative was the re-examination of the possibilities inherent in media understood as material (canvas, paper, wood, burlap, glass and so forth): see Schwitters and most Informal Art. But when absolute certainty, the unchanging values of the horizontal and vertical, of the primary values, were recognised, then the feeling arose that the approach adopted had been mistaken from the start and should be corrected. The error, that of identifying the eye as the instrument of understanding, derived from centuries of habit and mental sloth which restricted visual perception to Durer's maxim: "First, the eye that sees; second, the object that is seen; third, the intervening distance".

It was not enough to remove the object and the distance in order to achieve the optimum effect, known as abstraction. Malevitch, having recognised the law of geometrical absolution, was able to admit: "This is why we are obliged to clarify that – there is a difference – and indeed a great contradiction between the creative work of the painter and the pure optical activity." He continued: "Therefore we must conclude that the mutation of form and colour in artistic activity does not take place on the basis of visual optical perception, but as a consequence of their alterations in our minds; that is, in the creative imagination of the painter in whose mind the pictorial images has arisen."

The black square on the white square has a slight variation in measurement: it is an irregular square. The certainty of geometry has begun to yield and to transform itself into the Suprematist painting of 1917 – a geometric yellow form bathed in a light which dissolves into space. Certainty has by now completely given way to geometrical uncertainty, the conception of a geometrical form that is no longer determined by laws of mathematics or optics, but by slight dislocations and displacements produced by memory on the data of visual perception. Geometric irrationality has come into being as a result of the dissolution of geometrical form, its loss of weight, its escape from two-dimensionality: the instruments of enquiry have shifted decisively from perception to memory, together with all the knowledge, both conscious and unconscious, that the contemporary world has acquired. It is a shift so great as to produce a fundamental change in the visual world and in all future discoveries pertaining to it. It means "shaping imaginary space through a material object", as El Lissitsky foresaw.

Achille Perilli (1999)
Sussurrus Magicans

Achille Perilli (1999)
Il Tattile

What then is the imaginary space, if not a new model of visual structure, a more complex contribution to our definition of knowledge which is no longer linked to the activity of the eye alone, but related to a series of more complex sensations and different perceptions? And, with respect to history, does not this shift from the certainty of reason to the uncertainty of madness, reveal the decadence and dissolution of the utopia of technology, the safety of megalopolis, faith in progress? Vantongerloo during his last years followed Malevitch's example, substituting the mysticism of form with the mysticism of mathematics. The result was the name: the slow dissolution of geometrical certainty in favour of an open form permitting language to perceive the mysteries of the visual world through investigation.

In 1939 Variant seems to have extended its lines of force in space with an extraordinary essentiality of tension. For it is formal tension which causes the dissolution of form: it is tension which takes the place of order, certainty, safety.

Visual tension may only be achieved when several opposing forces are able to discharge themselves contemporaneously. When this takes place in a concentrated and *tense* space, such as geometrical space, the shifting of forces, the lessening of certainty – which interests me – then occurs. Form loses certainty and is transformed into a field of rapid motion, violent struggle, unbelievable deformations defining new, unknown and complex structures governed by laws which I defined in my 1975 manifesto *Machinerie, ma chere machine*:

From the relation between two geometric modules, occasionally in contrast, occasionally similar with slight variations, a sequence is born which tends to shift in space until ... it spreads from picture to picture, evolving, associating with other factors, constantly increasing the complexity of the investigation of *Imaginary Space*.

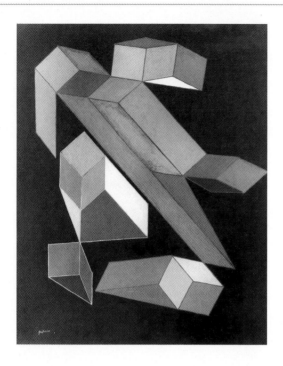

Achille Perilli (1999)
Il Signa Latente

42

The interpretation of such sequences suggests passages which shatter the idea of central space, or superficial space or light space. To enter within or to stay without, to advance or to halt, are all acts of vision, not of perception, which require independent choices for each of the various possibilities suggested by the image. It is not necessary to alter or change the order of sequence: alteration, transformation are inherent in the manner of seeing the structure, in the ambiguity cre-

Achille Perilli (1999)
Il Collasso del Carnale

ated by the shift form one level to another, from one chromatic law to another, from volume to void. Interpretation and observation must be extended in time to include successive stages of memorisation, to analyse possible interpretations without being limited to formal structures, but taking into account all these material belonging to the collective unconscious which inform the deepest levels of human consciousness."

It is a process which tends to broaden rather that restrict, to shift the field of enquiry from the perceptual to the mental; it rejects any minimilization of the uncertainties concentrated in the visual, expanding them so as to affect those uncharted areas between one code and another, and to implicate alien linguistic structures.

Such implications deny the severity of an absolute code, or a primary code and its unchangeable *linguistic* laws, since that code has already been transformed into *Another (Altro)* through a methodology which I call *intercode* : by working on areas which are affected by a law, but are not subject to it. Although these areas are sensitive to its presence and reality, the law perceived with the greatest ambiguity possible, with respect to meaning and value, so as to allow all manner of deformations and transformations. Only then is creative imagination able to build a new utopia: geometrical non-form.

References

[1] F. D'Amico, ed., *Achille Perilli 1968–1998: De Insana Geometria,* Comune di Ancona, Ancona (1998–1999).
[2] A. Perilli, *L'age d'or di Forma 1,* Corradini Editore, Mantova (1994).

43

Ovals in Borromini's Geometry

Michea Simona

The fourth centenary of the birth of the great Baroque architect Francesco Borromini (1599–1667) has seen many new studies on specific aspects of his work. They led to new considerations on the use of geometric forms in his architecture, confirmed by the recent architectonic relief of one of Borromini's most beautiful work: the church of San Carlo alle Quattro Fontane in Rome.

We concentrated in particular on Borromini's choice to use ovals according to the tradition of architecture treaties. Our aim is to give a first report in view of further researches on the subject.

The use of geometrical forms in Baroque architecture followed on the one side the tradition of the classical one, on the other it was founded on new ideas arisen in the Renaissance. Even if these forms are not so different from those used in previous periods, what changes is the way the architects deal with them. Ideal geometrical forms are not necessarily predetermined in the architectonic project: they become a means to plan buildings in which forms submit to constructive requirements and are consequently modifiable. The concept of geometric curve was for instance applied from the Greeks only to "classical" curves, while they refused to consider the curves drawn by mechanical way. On the contrary, the possibility to reduce general concepts to practical and technical solutions gained more importance in the 17th century.

At the beginning of the century the young Borromini spent his formative years in Milan, where there was a scientific milieu that brought him probably into contact with a new scientific culture for which the revival of ancient mathematics through translations was fundamental. An important contribute to these new ideas was given by two mathematicians of the circle of Urbino, representing the so called "mathematical humanism". The first of them, Federico Commandino (1509–1575), translated an important edition of the Conics of Apollonius of Perge (ca. 247–205 BC) published in 1566 and used by Johannes Kepler to determine the elliptical orbits of planets; the second one, Muzio Oddi (1569–1639), an architect living in Milan at the same time as Borromini, has to be considered a probable link between mathematics and architecture.

In architecture the oval shape was adopted as approximation of an ellipse, which was already studied in ancient times as one of the plane sections of a cone.

For a long time it was evident that an ellipse cannot be drawn directly with a pair of compasses, unless these are used to find a finite number of points of the ellipse, because it is a so called "polycentric" curve. This fact did not prevent several

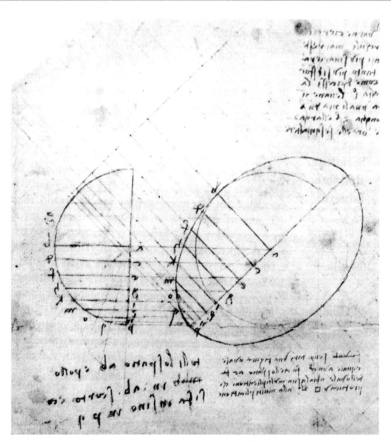

Fig. 1. Leonardo da Vinci, detail of the folio 318 b-r of the *Codex Atlanticus* (ca. 1510) with the construction of an ellipse as affine transformation of a circle. The construction seems to anticipate the axonometry of Eckhart (1926).

artists to draw ellipses already in the Renaissance: for example, Leonardo da Vinci presents in the *Codex Atlanticus* (ca. 1510) a construction of a true ellipse as affine transformation of a circle (Figure 1).

The question how to draw the shape of an ellipse in order to build elliptical forms in architecture was still studied in the Renaissance in architectural treaties, in particular in Il Primo Libro d'Architettura (1545) by Sebastiano Serlio (1475–1554), a fundamental work to which referred several architects in the following times.[1]

Serlio realised that the ellipse was a perfect shape being the representation of the circle in perspective, but he was also conscious how difficult it was to apply it in architecture, since neither geometric rules nor mechanical devices for drawing true ellipses are adequate to scale the draw up to accurate large ellipses. He said: "Many builders construct the ellipse with a thread, and then trace it with a pair of com-

---

[1] *Many concepts present in Serlio's work were already studied by his master Baldassarre Peruzzi (1481–1536). They were, however, left unpublished because of his early death.*

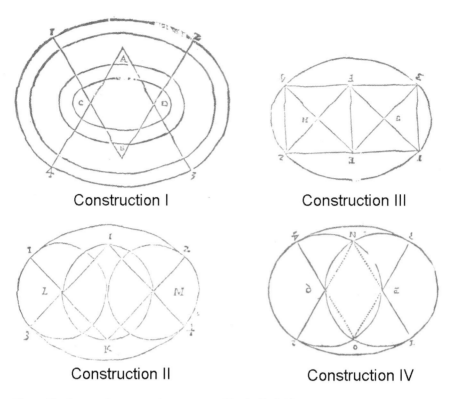

Construction I

Construction III

Construction II

Construction IV

47

Fig. 2. The four oval constructions proposed by Serlio [12].

passes, but this method does not work", in other words the "gardener's method" in which two pins are placed at the foci and a thread is attached to the pins (the sum of the distances from each point to the foci must be constant) is not easily applicable to buildings because of the imprecision of the method on large scale and the difficulty in applying it in the elevation of buildings. Another problem was to determine the perimeter of the ellipse: without complicated analytical solutions it can only be approximated, because of its continuing varying curvature.

Serlio suggested, therefore, in his First Book four solutions to approximate ellipses with ovals, i.e. shapes formed by four circular arcs. The advantage working with circles is evident when bisecting angles, constructing tangent and orthogonal lines, etc.

The question is nevertheless where to point the compasses to have ovals without tangent discontinuity in the connection points of the arcs. This problem was known and obviated using ovals with opportune entire ratios between radii of the circular arcs. It was however an open problem, solved only in recent times. For example, it can be shown that it is always possible to "close" two opposite circular arcs of same radius with other two arcs: circumscribing the two initial circles with a rectangle, there are always two arcs tangent to the medium points of the sides of the rectangle connecting the circumferences without tangent discontinuity [9].

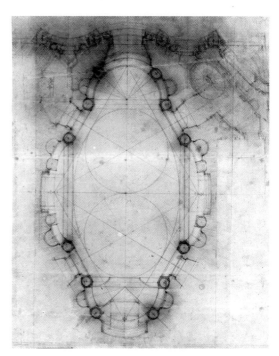

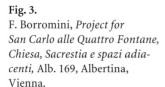

Fig. 3.
F. Borromini, *Project for San Carlo alle Quattro Fontane, Chiesa, Sacrestia e spazi adiacenti,* Alb. 169, Albertina, Vienna.

48

In each construction the oval is well determined by the four centres of the circular arcs, paired symmetric, which are the vertices of two congruent triangles with a common side. Among the four ovals proposed by Serlio, two were based on equilateral triangles, Construction I and IV, and two others on rectangle triangles (II and III).

Construction I can present different aspect ratios (between the semi-major and the semi-minor axes of the oval) [11], while Construction IV, particularly recommended by Serlio for its simplicity and therefore used as standard ellipse approximation in architectural practise [5], can also be seen as a particular case of Construction I.

In all the oval variants named above the basic figure of the triangle should not go farther than the circular arcs, i.e. with a vertex out of the oval. Giacomo Vignola (1507–1573) proposed a construction in which the height of the triangle (simply isosceles) coincides with the semi-minor axis and the circles determining the semi-major axis touch rather than intersect [5].[2] In this case the ratio between axis is 3:2 (and the one between radii of the circular arcs is 3:8) what guarantees tangent continuity: probably for this reason, it was already very used in Roman amphitheatres.

Borromini works with elementary geometrical constructions, but he distinguishes himself also for his continuous desire to introduce even remarkable

---

[2] *See for instance the dome of S. Andrea in Via Flaminia. For the use of oval in the Renaissance see [6].*

Fig. 4. The dome of San Carlo alle Quattro Fontane in Rome.

changes, often under the necessity of the advancing building's works, leading so to changes in the central geometrical constructions of the original draw.

With regard to the use of ovals he goes further introducing new variants: for example, the oval in the project for the ground-plan of the church of San Carlo alle Quattro Fontane in Rome (Figure 3)[3] is similar to the one of Vignola, but a vertex of each triangle is placed out of the oval. Moreover, the triangles are equilateral with the other two vertices in the centres of the circles defining the major axis. Doing so Borromini comes to a construction that has both equilateral triangles and touching circles dividing the plan in equal parts, finding a new variant of Serlio's Construction I. The ratio between the radii of the circular arcs is 3:1, and this is nevertheless an easy way to become tangent continuity in the connecting points (which is not longer guaranteed by varying the length of the semi-minor

---

[3] *Connors states that the geometric drawings for San Carlo alle Quattro Fontane have been made a posteriori in view of the publication of his complete work [4]. However, according to other drawings, the geometric construction has clearly not changed if compared with the original project.*

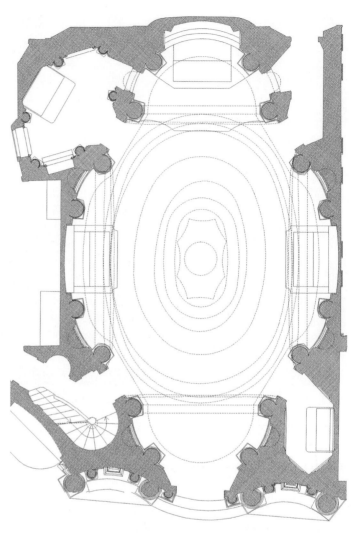

Fig. 5. San Carlo alle Quattro Fontane, Rome. The relief of the ground-plan with the projections of the dome: evidently, a direct relationship between the oval of the ground-plan and those of the dome is lacking. Relief by Alessandro Sartor, Rome.

axis)[4]. Besides, the circles result in their turn inscribed in larger equilateral triangles (with sides tangent in the connecting points of the oval) defining – with the above mentioned triangles – the partition of the church and the disposition of

---

[4] *The aspect ratio in this case is* $1 + \dfrac{1}{\sqrt{3}} \approx 1.5774$, *i.e. this oval is more stretched in comparison with each oval of Serlio*

(of course the Borromini's oval can be seen as a particular case of Serlio's Construction I, but Serlio did not seem to consider the variant with centres out of the oval).

niches and columns. This fact can be interpreted as a symbolical representation of a neo-platonic principle of tripartition, shown also by Steinberg in the use of geometrical elements in the church [13]. Many books at the beginning of the 17th century suggested a revival of neo-platonic terms – among them we mention the fundamental work of Johannes Kepler *Harmonices Mundi* – what lays emphasis on the peculiarity of Borromini, who joined together his skill as artisan and therefore his technical knowledge with a learned scientific culture, in agreement with the progressive diffusion of the new science. Nevertheless, the use of the equilateral triangle in the proportions is part of a middle-age geometrical tradition, with which Borromini probably came into contact in his youth, working at the Duomo in Milan. In another church of Borromini, S. Ivo alla Sapienza, the geometrical idea is based on an equilateral triangle too, even if everything in the building can induce to think of an hexagonal base: the triangle is hidden but always present.

It was Borromini's custom to extrapolate different solutions from the rule: the dome of S. Carlo has a complex geometrical internal surface requiring a partition of the oval base in eight equal sectors, and this is only possible with a canonical oval of the type of Serlio's Construction IV (whose arcs have the same length, what permits the partition). The problem is then to lean the dome on the structure based on the previously mentioned ground-plan oval: the recent architectonic relief emphasises a "correction" of the aspect ratio of the oval of the dome in order to match it to the lean points. This is part of Borromini's practice to submit the geometrical scheme to the requirements of elevations and surfaces. It is in particular possible for forms like the ovals, which permit a perfect interaction with the architecture and for which the artist's invention leads to new relationships between geometry and architecture.

Further investigations will concentrate on the way Borromini and other Baroque architects used ovals in order to adapt the ground-plan ovals to the Latin cross churches of previous periods. Some geometrical considerations seem to be necessary to conciliate the traditional forms with the Baroque innovative ones.

Acknowledgements

I would like to thank Nicola Soldini for his decisive suggestions resulting from his wide knowledge of Borromini's work.

Bibliography

[1] A. Bruschi, *Borromini: manierismo spaziale oltre il barocco,* Dedalo libri, Bari, 1978.

[2] AA.VV., *Galileo Galilei e gli scienziati del Ducato di Urbino,* Atti del convegno, Pesaro, 14 ottobre 1989.

[3] J. Connors, The First Three Minutes, in *Journal of the Soc. of Arch. Hist.,* 55 n.1 (1996), 38–57.

[4] J. Connors, Un teorema sacro: San Carlo alle Quattro Fontane, in *Il giovane Borromini. Dagli esordi a San Carlo alle Quattro Fontane,* Skira, Milano, 1999, 459–512.

[5] T. K. Kitao, *Circle and Oval in the Square of Saint Peter's,* New York University Press, 1974.

[6] W. Lotz, *L'architettura del Rinascimento,* Electa, Milano, 1997.

[7] A. Masotti, *Matematica e matematici nella storia di Milano,* estratto dal vol. XVI della *Storia di Milano* della Fondazione Treccani degli Alfieri per la storia di Milano, Milano, 1962.

[8] M. Oddi, *Degli Horologi Solari,* Ginammi, Venezia, 1638.

[9] F. Ragazzo, Geometria delle figure ovoidali, in *Disegnare,* 11 (1996), 17–24.

[10] M. Raspe, *Das Architektursystem Borrominis,* Deutscher Kunstverlag, München, 1994.

[11] P. L. Rosin, On Serlio's Constructions of Ovals, in *The Mathematical Intelligencer,* 23 n.1 (2001), 58–69.

[12] S. Serlio, *Tutte l'Opere d'Architettura,* de' Franceschi, Venezia, 1584.

[13] L. Steinberg, *Borromini's San Carlo Alle Quattro Fontane,* Garland Publishing, New York, 1977.

Francesco Borromini

Francesco Castelli, named Borromini, was born in 1599 in Bissone, in the southern part of Switzerland, and died in Rome in 1667. He was trained as a stone mason in the Milan cathedral, coming thus into contact with the Gothic architectural tradition. In 1619, he went to Rome, where he worked first in St Peter's workshop headed at that time by one of his relatives, Carlo Maderno. In Rome, Borromini studied the architectural works of the Antiquity and those of Michelangelo, from which he drew his inspiration.

After Maderno's death, Gian Lorenzo Bernini became Architect of St. Peter. Borromini worked for a time under Bernini, but only a few years later many incompatibilities divided the two architects: they became rivals.

Borromini spent the last years of his life in completing some unfinished building projects, under which some of his masterpieces: S. Ivo alla Sapienza, the interior of S. Giovanni in Laterano and his first work, San Carlo alle Quattro Fontane.

The fourth centenary of Borromini's birth led to many celebrations. In particular, three exhibitions were hold in Lugano ("Il giovane Borromini", Museo Cantonale d'Arte, September – December 1999), in Rome ("Il Borromini e l'universo Barocco", Palazzo delle Esposizioni, December 1999 – February 2000) and in Vienna ("Francesco Borromini – Struktur und Metamorphose", Graphische Sammlung Albertina, April–May 2000). An international congress took place in Rome in January 2000.[5]

The Accademia di architettura of the Università della Svizzera italiana, on that occasion, gave the commission to do the relief of one of Borromini's most important works: the Church of San Carlo alle Quattro Fontane.[6]

---

[5] *Catalogues of the exhibitions and the congress:*
M. Kahn-Rossi, M. Franciolli, eds., Il giovane Borromini. Dagli esordi a San Carlo alle Quattro Fontane (exhibition held at the Museo Cantonale d'Arte in Lugano), Skira, Milano, 1999.
R. Bösel and C. Frommel, eds., Borromini e l'universo barocco (exhibition held at the Palazzo delle Esposizioni in Rome), 2 vol., Electa, Milano, 1999.
R. Bösel, ed., Borromini. Architekt im barocken Rom, (exhibition held at the Albertina in Vienna), Electa, Milano, 2000.
Francesco Borromini, Atti del Convegno internazionale, Roma, gennaio 2000, Electa, Milano, 2000.

[6] *The relief will be published by N. Soldini and A. Sartor.*

Fractals: A Resonance between Art and Nature

RICHARD TAYLOR, BEN NEWELL, BRANKA SPEHAR and COLIN CLIFFORD

Physics and Psychology Reveal the Fractal Secrets of Jackson Pollock's Drip Paintings

The discovery of fractal patterns was an interesting advance in the understanding of nature [1, 2]. Since the 1970s many natural scenes have been shown to be composed of fractal patterns. Examples include coastlines, clouds, lightning, trees, rivers and mountains. Fractal patterns are referred to as a new geometry because they look nothing like the more traditional shapes such as triangles and squares known within mathematics as Euclidean geometry. Whereas these shapes are composed of smooth lines, fractals are built from patterns that recur at finer and finer magnifications, generating shapes of immense complexity. Even the most common of nature's fractal objects, such as the tree shown in Figure 1, contrast sharply with the simplicity of artificially constructed objects such as buildings. But do people find such complexity visually appealing? In particular, given people's continuous visual exposure to nature's fractals, do we possess a fundamental appreciation of these patterns – an affinity independent of conscious deliberation?

The study of human aesthetic judgement of fractal patterns constitutes a relatively new research field within perception psychology. Only recently has research started to quantify people's visual preferences for (or against) fractal content. A useful starting point in assessing people's ability to recognize and create visual patterns is to examine the methods used by artists to generate aesthetically pleasing images on their canvases. More specifically, in terms of exploring an intrinsic appreciation of certain patterns, it seems appropriate to examine the Surrealists and their desire to paint images which are free of conscious consideration. Originating in Paris during the 1920s, the Surrealists developed their painting techniques more than fifty years ahead of the scientific discovery of nature's underlying fractal quality. Yet, remarkably, our recent research shows that fractals could have served as the foundation for Surrealist art and, in particular, the drip paintings of their artistic offspring, the American abstract painter Jackson Pollock.

The Surrealists' approach to painting deviated radically from the care and precision traditionally associated with artistic techniques. The Surrealists believed that premeditated, conscious actions hindered the liberation of *pure* imagery from deep within the mind [3]. They thought that the key to releasing this imag-

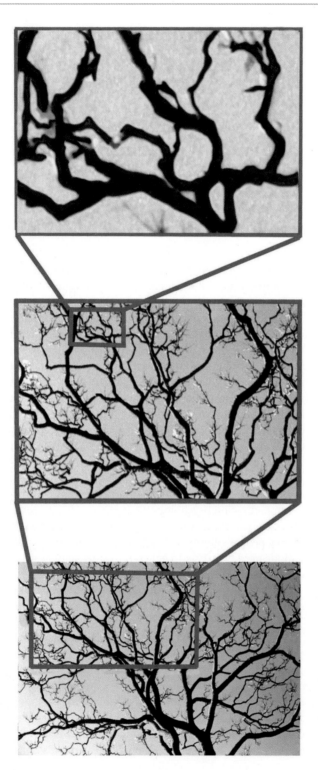

Fig. 1. Trees are an example of a natural fractal object. Although the patterns observed at different magnifications don't repeat exactly, analysis shows them to have the same statistical qualities (photograph by R.P. Taylor)

Fig. 2. A photograph of peeling wall paint. As a demonstration of *free association*, a picture of a person (*left*) and a bird (*right*) have been drawn based on images perceived within the peeling paint (photograph by R.P. Taylor)

55

ery lay in the exploitation of chance happenings. By staring at random patterns, such as those produced by a spilled bottle of ink, a chance arrangement of patterns might trigger the imagination and cause an image to emerge in the artist's mind. Adopting the Surrealist terminology, the random patterns were said to serve as a *springboard for free association*. The artist would then draw over the top of this springboard pattern, building a picture based on this initial perceived image. Interestingly, Leonardo da Vinci had already suggested a similar approach back in 1500 in his *Treatise On Painting*: "… a new inventive kind of looking consists in this, that you look at a wall which is marked with all kinds of stains. If you have to invent a situation, you can see things in it that look like various landscapes. Through confused and vague things the spirit wakes to new inventions." The technique is demonstrated in Figure Two, where we have drawn pictures of a person and a bird within the patterns of peeling wall paint. Whereas da Vinci's approach was passive – he simply used the springboards in his surrounding environment – the Surrealists actively created their own springboards by generating what they regarded as random patterns.

André Masson threw handfuls of sand over a canvas covered with glue. On tilting the canvas, sand fell off in some regions but not others. He then painted a human figure based on his perceptions of the springboard pattern of sand. Oscar Dominguez invented a technique labeled as *decalcomania*, where paint was spread on a sheet of paper, another sheet was pressed down lightly on top and then was lifted off before the paint had dried. He described the resulting pattern as "unequalled in its power of suggestion." Similarly, Joan Miró spread diluted paint across a canvas using a sponge in a random manner to encourage a chance emergence of a pattern. In 1925, Max Ernst introduced his *frottage* technique, where the springboard was created by randomly placing sheets of paper on the surface of an old wooden floor and taking rubbings with black lead. Ernst regarded this as a major breakthrough, remarking, "in gazing intently at the drawings thus obtained I was surprised by the sudden intensification of my visionary capacities". By the 1940s, Ernst had moved on to a new technique where paint was dripped onto a horizontal canvas from a leaking can swung randomly through the air on a piece of string. He moved to New York and stimulated a new generation of artists who later were to become known as the Abstract Expressionists. The most famous of these was Jackson Pollock, who, similar to Ernst, dripped paint from a can onto large horizontal canvases. Acknowledging the strong influence of the Surrealists, Pollock noted, "I am particularly impressed with their concept of the source of art being the unconscious." During Pollock's artistic peak of 1940s–1950s, however, art critics were generally unsympathetic to his achievements, describing his work as "mere unorganized explosions of random energy, and therefore meaningless." [4]

Whereas the Surrealists and their artistic offspring, the Abstract Expressionists, used these random patterns to trigger imagery for their artistic creations, the psychologists of the same era used similar patterns in the hope of assessing people's mental and emotional disorders. The most famous examples of this are the ink blot psychology tests introduced by Hermann Rorschach [5]. Rorschach's technique was inspired by a popular children's game known as blotto, where the players were asked to identify images within the patterns created by ink blots. Rorschach developed this simple concept into his Form Detection Tests, where the blot patterns were thought to act as springboards for free association and the images perceived by the observer were interpreted as direct projections of the unconscious mind. Rorschach died in 1922 having devoted just four years to his ink blot tests. However, during the 1940s and 1950s, the *Rorschach*, as it became known, became the test of choice in clinical psychology for assessment of mental disorders. While the use of these patterns for mental assessment is now only of historical interest, the ink blots clearly evoke meaningful images (Figure 3).

A recent perception study of free association has triggered renewed interest in the fundamental characteristics of ink blot patterns [6]. Bernice Rogowitz and Richard Voss investigated people's responses to fractal patterns. To do this they quantified the fractal patterns' visual character using a parameter called the fractal dimension, D. This parameter describes how the patterns occurring at different magnifications combine to build the resulting fractal pattern. As the name fractal dimension suggests, this building process determines the dimension of the fractal pattern. For Euclidean shapes, dimension is a simple concept and is de-

Fig. 3. Ink blot patterns created by R.P. Taylor using the technique employed by Rorschach when generating his ten original patterns

scribed by familiar integer values – for a smooth line (containing no fractal structure) D has a value of 1, whilst for a completely filled area (again containing no fractal structure) its value is 2. However, the repeating structure of a fractal pattern causes the line to begin to occupy area. D then lies between 1 and 2 and, as the complexity and richness of the repeating structure increases, its value moves closer to 2. Figure 4 demonstrates how a fractal pattern's D value has a profound effect on its visual appearance. For fractals described by a low D value close to one (left), the patterns observed at different magnifications repeat in a way that builds a very smooth, sparse shape. However, for fractals described by a D value closer to two the repeating patterns build a shape full of intricate, detailed structure (right).

The research by Rogowitz and Voss indicates that people perceive imaginary objects (such as human figures, faces, animals etc.) in fractal patterns characterized by low D values [6]. For fractal patterns with increasingly high D values this perception falls off markedly. This result caused Rogowitz and Voss to speculate that the ink blots used to induce projective imagery in psychology tests of the 1920s were fractal patterns described by low D values. Indeed, their subsequent

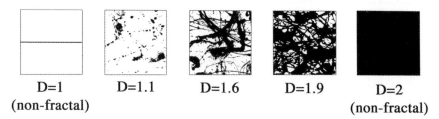

D=1
(non-fractal)   D=1.1   D=1.6   D=1.9   D=2
(non-fractal)

Fig. 4. A comparison of patterns with different D values: 1 (*left*), 1.1, 1.6, 1.9 and 2 (*right*)

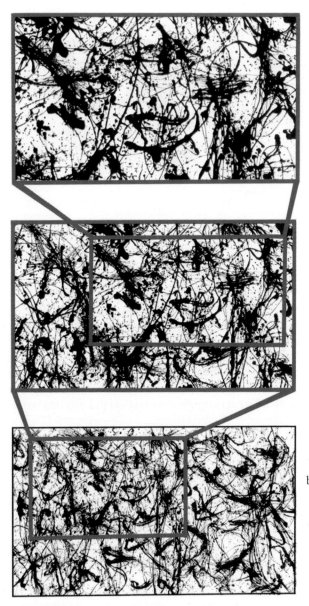

58

Fig. 5. Pollock's *Number 32,1950* reveals patterns at different magnifications. The fractal content of Pollock's dripped paintings has been confirmed by pattern analysis techniques. *Number 32, 1950* (enamel on canvas, 269 by 457.5cm) was painted by Pollock in 1950 (Kunstsammlung Norrhein-Westfalen, Düsseldorf)

preliminary analysis indicated that ink blots were fractal with a D value close to 1.25. Perhaps, then, the springboard patterns produced by the Surrealists, and later by the Abstract Expressionists, were also fractal? The repeating quality of Jackson Pollock's dripped paintings at different magnifications (shown in Figure 5) supports this speculation and our recent analysis of his work confirms their fractal content [7].

During his *classic* period of 1950, Pollock was filmed whilst painting. This serves as a remarkable visual record of how he used his perfected drip technique to build his fractal patterns. Our analysis of this film reveals that he differed from his Surrealist forerunners in one crucial respect. After 20 seconds of the dripping process, Pollock had established a fractal pattern with a low D value [2]. The Surrealists (and clinical psychologists) would have stopped at this initial stage and then used the pattern as a springboard for free association. For example, Max Ernst stopped dripping paint at the equivalent stage of his painting process and stared at the springboard layer in the hope of perceiving an image within the swirls of dripped paint. Then, based on this perception, Ernst drew a picture on top of the springboard layer. For the Surrealists, this drawing process was often so heavy that it obscured the underlying springboard layer, making a fractal analysis of this layer difficult. In contrast to this Surrealist technique, Pollock didn't stop dripping paint once the low D fractal springboard pattern had been established. Instead, he continued to drip paint for a period lasting up to six months. Depositing layer upon layer, he gradually built a highly dense fractal pattern. As a result, the D value of his paintings rose gradually as they neared completion, starting in the range of 1.3 to 1.5 for the initial springboard layer and reaching a final value as high as 1.9 [2].

When combined with the findings of Rogowitz and Voss, this time-sequence analysis provides an answer to one of the more controversial issues surrounding Pollocks drip work. Over the last 50 years there has been a persistent theory that speculates that Pollock painted illustrations of objects during the early stages of the painting's evolution and then deliberately obscured them with subsequent layers of paint [8]. In reality, the low D values evident during the early stages of the painting process simply caused the observer to perceive objects in the dripped patterns (even though they were not there) and these perceptions were then suppressed (making the objects apparently disappear) as D rose to the high value which characterized the complete pattern ([2].

Pollock's desire to paint fractal patterns is not surprising. Our initial perception studies revealed that over ninety percent of 120 participants found fractal imagery to be more visually appealing than non-fractal imagery [9, 10]. However, it is clear from our film analysis that Pollock's painting process was geared to more than simply generating a fractal pattern – if this were the case he could have stopped after twenty seconds, having established his fractal springboard pattern. Instead he invested a further six months fine-tuning his pattern to produce a fractal painting described by a high D value. Furthermore, his ability to paint fractal patterns with such high D values represented the culmination of almost ten years work during which he steadily perfected his drip process. When he first started to drip paint in 1943, he didn't build beyond the initial springboard layer. Inheriting the Surrealist technique, he used the low D fractal patterns to evoke images and then named his paintings after these images (*Eyes in the Heat* and *Water Birds* are example titles). In contrast, many of his paintings from his classic period of 1950 to 1952 were simply numbered or left untitled, presumably because the high D values of these fractal patterns no longer evoked any images. So why would Pollock invest so much effort in creating fractal patterns with such high D values? Perhaps he found such patterns to be aesthetically pleasing?

Fig. 6. Clouds form fractal patterns which, according to our survey, are aesthetically pleasing (photograph by R.P. Taylor)

60

In 1995, Cliff Pickover used a computer to generate fractal patterns with different D values and found that people expressed a preference for fractal patterns with a high value of 1.8 [11], similar to Pollock's paintings. However, a survey by Deborah Aks and Julien Sprott also used a computer but with a different mathematical method for generating the fractals. This survey reported much lower preferred values of 1.3 [12]. Aks and Sprott noted that the preferred value of 1.3 revealed by their survey corresponds to fractals frequently found in natural environments (for example, clouds and coastlines have this value) and suggested that perhaps people's preference is actually *set* at 1.3 through a continuous visual exposure to nature's patterns. However, the discrepancy between the two surveys seemed to suggest that there is not a universally preferred D value but that the aesthetic qualities of fractals instead depend specifically on how the fractals are generated. There are, in fact, three fundamentally different ways in which fractals can be generated – by nature's processes, by mathematics and by humans (as revealed by our analysis of Pollock's paintings). To determine if there are any *universal* aesthetic qualities of fractals, we therefore carried out a survey incorporating all three categories of fractal pattern and found that – irrespective of their origin – there was a distinct preference for D values in the range 1.3 to 1.5 [13]. Figure 6 shows an example that the survey revealed to be aesthetically pleasing – clouds with a D value of 1.3.

Perception studies of fractal patterns clearly have wide ranging implications for the types of environment which people find fundamentally pleasing. Our results indicate, for example, that architects should consider incorporating low D fractals into the interior and exterior surfaces of future building designs. Indeed, a dramatic demonstration of this occurred in November 2000 when the Guggenheim Museum unveiled plans for a new $ 800M building to house its modern art collection in New York [13]. Composed of swirling layers of curved surfaces, the 45 story structure is designed by the architect Frank Gehry to be cloud-like and is expected to radically re-shape Manhattan's waterfront. Although Gehry's building proposal for the Guggenheim Museum is designed to mimic the general form of clouds, it is clear that the completed building will not strictly be fractal. To build a structure described by a D value of 1.3 would require many layers of repeating patterns. Although this is no great challenge for nature, such complexity is beyond current building techniques. In fact, both Gehry and New York's former major, Rudolph Giuliani, readily admit that no shovel will be turned for at least 5 years and that the plans will have to evolve between now and then. However, it will be fascinating to see if people's fundamental appreciation of fractal clouds will inspire New Yorkers to embrace this revolutionary building design.

As for Jackson Pollock, he remains an artistic enigma. He could have stopped his painting process after less than a minute, having generated a pattern with a relatively low D value that people would have found visually appealing. Instead he spent six months depositing further layers, evolving the painting towards a higher D composition and apparently away from the aesthetic ideal. Should we conclude that Pollock wanted his work to be aesthetically challenging to the gallery audience? Perhaps Pollock's artistic achievement lies in his rebellion against our fundamental affinity for low D fractals. James Wise recently speculated that humans find low D fractal patterns aesthetically pleasing because these patterns make us feel safe in terms of our survival instincts. For example, it is easier to detect predators within natural scenery composed of sparse structure (low D fractals) than complex structure (high D fractals) [14]. Alternatively, as discussed above, Aks and Sprott argue that our preference for low D patterns occurs simply because these patterns are more abundant in nature and that we acquire (either implicitly during our lifetimes or through evolution) an appreciation for what we are familiar with [12]. Which ever of these two theories we apply to Pollock's paintings, the low D patterns painted in his earlier years should have a more calming effect than his later *classic* drip paintings. What was motivating Pollock to paint high D fractals? It is possible that he regarded the restful experience of a low D pattern as being too bland for an artwork and wanted to keep the viewer alert by engaging their eyes in a constant search through the dense structure of a high D pattern. We plan to investigate this intriguing possibility by performing eye-tracking experiments on Pollock's paintings, which will assess the way people visually assimilate fractal patterns with different D values.

In light of Pollock's interest in Surrealist techniques – the art movement most closely associated with operations of the mind – it is fitting that the discipline providing recent insights into the visual significance of Pollock's work is that of psychology. The impact of Pollock's work on psychology research extends beyond our perception studies of his fractal patterns. Whereas perception psychologists are interested in the visual impact of Pollock's completed patterns, behavioral psychologists are intrigued by his painting process and how a human was able to generate fractal patterns. Recently, we described Pollock's style as Fractal Expressionism to distinguish it from computer-generated fractal art [15]. Fractal Expressionism indicates an ability to generate and manipulate fractal patterns *directly*. Furthermore, our analysis shows that the fractal quality of Pollock's paintings was established within a remarkably quick time frame – within less than one minute. How could someone paint such intricate and complex fractal patterns, so precisely, so quickly and do so 25 years ahead of the scientific discovery of fractals?

A common interpretation of Pollock's painting process focuses on the Surrealist technique called *psychic automatism* [3]. In this technique, artists painted rapidly and spontaneously, with such speed that conscious intervention was thought to be suppressed. In this way, their gestures were regarded as being *steered* by the unconscious. Critics have since questioned whether psychic automatism can be achieved in reality. In 1959, Rudolph Arnheim rejected the concept as "romantic" and proposed that the relaxation of conscious control would simply lead to nothing more than a confused and patternless disorder [16]. However, as we have seen, the patterns pouring onto Pollock's canvas weren't disorganized – they were fractal. Why would this be? Questions such as this have attracted the attention of medical researchers who investigate the basic rhythms of the human body. Ary Goldberger and his research team study the dynamics of human processes that operate independently of conscious control, including heart beats and stride length during walking [17]. They conclude that strict periodicity in such processes is a signature of a pathological condition and that a healthy behavior instead reveals fractal variations around this periodicity. This suggests that, in surrendering conscious control, the Surrealist method of automatism might tune into the basic fractal rhythms of the human body and that Pollock applied this to his drip technique. Future research will be needed to further explore the link between the aesthetic qualities of fractal patterns and the human ability to paint them.

Acknowledgments

We thank Adam Micolich (Physics Department, University of Oregon, Eugene, 97403-1274, USA) and David Jonas (Physics Department, University of New South Wales, Sydney, 2052, Australia).

References

[1] B.B. Mandelbrot, *The Fractal Geometry of Nature*, New York, W.H. Freeman and Company, 1977 and *The Visual Mind*, M. Emmer, ed., MIT Press, 1993.

[2] R.P. Taylor, A.P. Micolich and D. Jonas, "The Construction of Jackson Pollock's Fractal Drip Paintings", to be published in *Leonardo*, 35, 202, 2002.

[3] André Breton outlined the aims of Surrealism in his *Manifeste du Surréalisme*, Paris 1924. See, for example, "Dada and Surrealism" by Dawn Ades in *Concepts of Modern Art*, ed. N. Stangos, London, Thames and Hudson, 1994.

[4] E.G. Landau, *Jackson Pollock*, London, Thames and Hudson, 1988.

[5] H. Rorschach, *Psychodiagnostics*, New York, Grune and Straton, 1921.

[6] R.E. Rogowitz and R.F. Voss, "Shape Perception and Low Dimension Fractal Boundary Contours", *Proceedings of the Conference on Human Vision: Methods, Models and Applications*, S.P.I.E., **1249**, 387, 1990.

[7] R.P. Taylor, A.P. Micolich and D. Jonas, "Fractal Analysis of Pollock's Drip Paintings", *Nature*, **399**, 422, 1999.

[8] K. Varnedoe and P. Karmel, *Jackson Pollock*, New York, Abrams, 1998.

[9] R.P. Taylor, "Splashdown", *New Scientist*, **2144**, 30, 1998.

[10] R.P. Taylor, A.P. Micolich and D. Jonas, "The Use of Science to Investigate Jackson Pollock's Drip Paintings", *Journal of Consciousness Studies*, **7,** 137, 2000.

[11] C. Pickover, *Keys to Infinity*, New York, Wiley, 206, 1995.

[12] D. Aks and J. Sprott, "Quantifying Aesthetic Preference for Chaotic Patterns", *Empirical Studies of the Arts*, **14,** 1, 1996.

[13] R.P. Taylor, "Architect Reaches for the Clouds", *Nature*, **410**, 18, 2001.

[14] See, for example, J. A. Wise and T. Leigh Hazzard, "Bionomic Design", *Architech*, 24, Jan. 2000.

[15] R.P. Taylor, A.P. Micolich and D. Jonas, "Fractal Expressionism", *Physics World*, **12,** 25–28, 1999.

[16] R. Arnheim, "Accident and the Necessity of Art", *Journal of Aesthetics and Art Criticism*, **16,** 18, 1959.

[17] J.M. Hausdorff, P.L. Purdon, C.K. Peng, Z. Ladin, J.Y. Wei and A.L. Goldberger, "Fractal Dynamics of Human Gait: Stability of Long Range Correlations in Stride Interval Fluctuations", *Journal of Applied Physiology*, **80,** 1448, 1996 and A.L. Goldberger, "Fractal Variability Versus Pathologic Periodicity: Complexity Loss and Stereotypy in Disease", *Perspectives in Biology and Medicine*, **40,** 543 Summer edition 1997.

63

Mathland – From Topology to Virtual Architecture

Michele Emmer

In mathematics lies the essence of the spirit
Robert Musil

Premise

In the summer of 2002 the Architecture Biennale was held in Venice. Among the many projects and ideas on show, some very interesting, others merely more or less extravagant, was the project for a museum of the Hellenic world from a group of architects called *Anamorphosis Architects,* comprised of Nikos Georgiadis, Tota Mamalaki, Kostas Kakoyiannis and Vaios Zitounolis.

Fig. 1.
*Museum
of the Hellenic World,*
Anamorphosis
architects,
Athens, Greece.
Northeastern view
from the piazza.
© Anamorphosis
architects

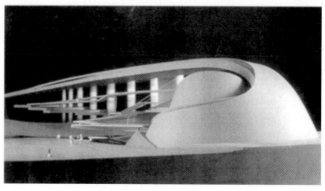

*Museum
of the Hellenic World,*
Anamorphosis
architects,
Athens, Greece.
Northeastern view,
model with the roof
removed.
© Anamorphosis
architects

A project that places a great emphasis on the spatiality of the construction, it has a large open and continuously evolving space, with those curved lines that twist into contorted spirals, and in the centre, right at the centre of a great spiral, the expository heart of the classic period of Greek civilisation. That building was, in some sense, the start and (temporary) end of a discourse that began with Euclidean geometry thousands of years ago. It was a geometry that, together with Greek philosophy, was at the heart of the foundation of western civilisation, as we now know it. Obviously, let us not forget the influence of lots of other civilisations, above all the Islamic one, which has allowed Europe to rediscover the forgotten Greek civilisation.

There are several questions that must be investigated in order to understand, at least partly, how philosophical, artistic, scientific and, in a word, cultural elements have contributed throughout the centuries to the synthesis of a project such as that for Hellenic civilisation. A sort of voyage to the inside of western culture of the past two millennia and more, highlighting from my point of view the cultural aspects associated to geometry, mathematics and architecture.

Space and Mathematics

"I feel I am perceiving a decline in the belief that in philosophy it is necessary to lean on the opinions of some celebrated author; as if our mind should remain completely sterile and infertile when not mixed with another's discourse; and perhaps it should believe philosophy is a book or the fantasy of man, such as *The Iliad* or *Orlando Furioso*, books in which the least important thing is that what is written is true. But things are not thus. Philosophy is written in this great book that is continuously open in front of our eyes (I mean the universe), but it cannot be understood without first learning to understand its language and characters, and what is written. It is written in a mathematical language and the characters are triangles, circles and other geometric figures; without these it is impossible to humanly understand a word; without these it is mere wandering in vain around a dark maze."

Words of Galileo Galilei, written in *The Assayer*, published in Rome in 1623. Without mathematical structures one cannot understand nature. Mathematics is the language of nature. We now leap forward many centuries.

In 1904 a famous painter wrote as follows to Emile Bernard: "Traiter la nature par le cylindre, la sphère, le cône, le tous mis en perspective, soit que chaque côté d'un objet, d'un plan, se dirige vers un point central. Les lignes parallèles à l'horizon donnent l'étendue, soit une section de la nature ... Les lignes perpendiculaires à cet horizon donnent le profondeur. Or, la nature, pour nous hommes, est plus en profondeur qu'en surface, d'où la nécessité d'introduire dans nos vibrations de lumière, représentée par les rouges et le jaunes, une somme suffisante de bleutés, pour faire sentir l'air."

The historian of art, Lionello Venturi commented that you don't see any cylinders, spheres or cones in Cézanne's paintings, so the sentence merely expressed an ideal aspiration to an arrangement of shapes that transcend nature, nothing else.

In the same years as Cézanne was painting and even a little earlier, the shape of geometry was changed from that of Galileo's day. Geometry was profoundly altered in the course of the second half of the 19th Century. In the years between 1830 and 1850, Lobacevskij and Bolyai constructed the first examples of non-Euclidean geometry, in which Euclid's famous fifth postulate on parallel lines was not valid. Not without doubts and deliberations, Lobacevskij was to call his geometry *imaginary geometry* (nowadays called non-Euclidean hyperbolic geometry), it was so counter to common sense. Non-Euclidean geometry remained a marginal aspect of geometry for several years, a sort of curiosity, until it was finally incorporated into mathematics as an integral part via the far-reaching ideas of G.F.B. Riemann (1826–1866). In 1854 Riemann brought before the Faculty of the University of Göttingen the famous dissertation entitled *Ueber die Hypothesen, welche der Geometrie zu Grunde liegen* (On the Hypotheses which lie at the Bases of Geometry), that was to be published only in 1867. In his presentation, Riemann upheld a global vision of geometry as the study of varieties in any number of dimensions, in any kind of space. According to Riemann's ideas, geometry need not necessarily deal with points or space in the ordinary sense, but sets of n-ple coordinates. On becoming a professor at Erlangen in 1872, Felix Klein (1849–1925) in his inaugural speech, known as the *Erlangen Program*, described geometry as the study of the properties of shapes that have an invariant character with respect to a particular group of transformations. Consequently every classification of the groups of transformations becomes an encoding of the different geometries. For example, the Euclidean geometry of the plane is the study of the properties of shapes that remain invariant with respect to the group of rigid transformations of the plane, comprised of the set of translations and rotations. Jules Henri Poincaré states that "the geometric axioms are neither synthetic a priori rules, nor experimental facts. They are conventions: our choice, among all possible conventions, is guided by experimental facts, but remains free and unlimited by the need to avoid all contradictions. This is how the postulates can remain rigorously true, even if the experimental laws that have determined their selection are nothing but approximations. In other words, the axioms of geometry are nothing but dressed up definitions. Therefore, why ponder the question 'Is Euclidean geometry the true one?' It makes no sense. Just as it makes no sense to ask if the metric system is the true one and the old measurement systems false. One geometry cannot be more valid than another; it can only be more convenient. Euclidean geometry is, and always will be, the most convenient one."

The official birth of that branch of mathematics now called *Topology* is also due to Poincaré, in his book *Analysis Sitûs*, a Latin translation of a Greek name, published in 1895: "As regards me, all the different research I have undertaken has led me to *Analysis Sitûs* (literally *Analysis of Position*)." Poincaré defined Topology as the science of understanding the qualitative properties of geometric shapes not just in ordinary space but also in space of dimensions greater than three.

Adding to all this the geometry of complex systems, the geometry of fractals, chaos theory and all the 'mathematical' images discovered (or invented) by mathematicians in the last thirty years using computer graphics, we can easily understand how mathematics has contributed crucially to changing our notion of space many times over, both the space we live in and the notion of space itself. We see that mathematics is not a mere kitchen utensil, but has contributed to, and sometimes determined, the way in which we understand space on earth and in the universe.

This is particularly so as regards Topology, the science of transformations and the science of invariants. See for example Frank O. Gehry's project in New York for the new Guggenheim museum in Manhattan, an even more provocative, even more topological project than that of the Guggenheim in Bilbao.

The cultural leap is certainly remarkable; building with technologies and materials that allow us to effect a transformation, making the building nearly continuous, a kind of contradiction between the final construction and its deformation. It is an interesting sign that one has started to study contemporary architecture using the tools that mathematics, that science has provided, cultural tools as well as technical ones.

It is worthwhile highlighting how the discovery (or invention) of non-Euclidean and higher dimensional geometries, starting with dimension four, is also one of the most interesting examples of the profound repercussions that many mathematicians' ideas were to have on the culture of humanities and the arts.

Like every good journey, we need to trace a route, a route that will include the elements used to make sense of the word *space*.

The first element is, without shadow of a doubt, the space that Euclid came to outline with definitions, axioms and the properties of objects that were to lie in that space. The space that would be the one of perfection, Platonic space. Man as origin and measure of the universe, a notion that transcends centuries. The mathematics, the geometry that has to explain everything, even the shape of living things: *The Curves of Nature*, the title of a famous 20th Century book by Cook, which one certainly could not have imagined to be true, finding in the shapes of nature as it does, even in those right at the origin of life, several mathematical curves. From the famous book by D'Arcy Thompson, *On Growth and Form* of 1914 to René Thom's catastrophe theory, to complexity and the Lorentz effect and non-linear dynamical systems.

The second element is liberty; mathematics, geometry would seem to be the kingdom of dull. Anyone who has never practiced mathematics, who has never studied mathematics at school with interest, will never manage to understand the deep emotion that mathematics can inspire. Nor can these people understand that mathematics is a highly creative activity. Nor that it is the kingdom of liberty where not only do you invent (or discover) new objects, new theories, new active areas of research, but you also invent the problems. Moreover, since in many cases mathematics does not require huge financial resources, one can rightly say that mathematics is the kingdom of liberty and fantasy. And certainly of rigour. And correct reasoning.

The third element on which to reflect is how all these ideas are communicated and assimilated, hopefully not buried deep and only heard of by different sectors of society. The architect Alicia Imperiale wrote in the chapter *Digital Technologies and New Surfaces* of the book *New Bidimensionalities* [1]: "Architects freely appropriate specific methodologies from other disciplines. This can be attributed

Fig. 2

Fig. 3

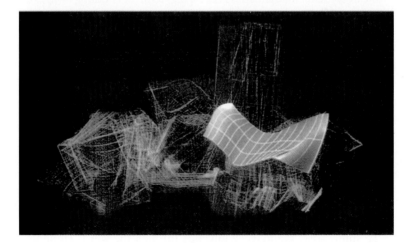

Fig. 4

Fig. 2–4. Frank O. Gehry, *New Guggenheim Museum,* New York
© courtesy of Keith Mendenhall for the GEHRY PARTNERS studio

to the fact that broad cultural changes are verified quicker in other contexts than in architecture." She adds:

> Architecture reflects the changes that occur in culture, albeit many feel at a painfully slow pace. [...] In constantly seeking to occupy an avant-garde role, architects think the information borrowed from other disciplines can be rapidly assimilated into architectural design. Nevertheless, this 'translatability', the transfer from one language to another, remains a problem. [...] Architects more and more frequently look to other disciplines and industrial processes for inspiration, and make ever greater use of computer design and industrial production software originally developed for other sectors.

Later on, Ms Imperiale recalls, "it is interesting to note that, in the information era, disciplines once distinct are now linked to each other through a universal language: the digital binary code." Does the computer resolve all problems?

The fourth element is the computer, the graphic computer, the logic machine and geometer par excellence. The idea of an intelligent machine realised, a machine capable of attacking the widest range of problems, if we are able to make it understand the language we are using. The inspired idea of a mathematician, Alan Turing, [2] brought to fruition by stimulus of a war. A machine built by man, into which has been inserted a logic also built by man, thought up by man. A very sophisticated tool, irreplaceable, not just for architecture. Precisely that, a tool.

The fifth element is progress, the word progress. If one considers non-Euclidean geometries, new dimensions, Topology, the explosion of geometry and mathematics in the Twentieth Century, can one talk of progress? Of knowledge, certainly, but not in the sense that the new results cancel the preceding ones. Mathematicians are prone to say that "mathematics is like a pig, it throws nothing away, and sooner or later even the things that seem most abstract and senseless can become useful." Imperiale writes that Topology is effectively an integral part of the field of Euclidean geometry. What has escaped the writer of these words is, what the word *space* means in geometry. Just words. Where instead a change of geometry helps tackle problems that are different just because the structure of the space is different. Space is the properties, not the objects contained therein. Words.

Words are the sixth element. One of the greatest skills of humanity is the ability to give names to things. Often when naming one uses words that are already in use. This practice can sometimes cause problems, because on hearing the word one has the impression of understanding, or at least gleaming a sense of, what it means. This has happened, especially in recent years, in mathematics with words like fractals, catastrophes, complexity, hyperspace. Symbolic, metaphoric words. Topology, dimensionality and serialisation have also now entered into common usage, at least among architects.

To summarise, the journey unwinds through the words computers, axioms, transformations, words, liberty. One word will have a great importance in this journey for the idea of space: Topology. For the other aspects I refer to the book *Mathland: from Flatland to Hypersurfaces* [3].

From Topology to Virtual Architecture

"Around the middle of the 19th century, geometry began to develop in a completely different way that was destined to quickly become one of the great forces of modern mathematics."

The words of Courant and Robbins in their famous book, *What Is Mathematics.* "This new field, called *analysis situ* or topology, has as its object the study of the property of geometric figures that persist even when the figures undergo deformations so profound as to lose all their metric and projective characteristics."

Topology, then, has as its subject the study of those properties of geometric shapes that remain invariant when the shape is subjected to deformations so complete as to lose all their metric and projective properties, such as shape and dimension. That is, the geometric shapes retain only their qualitative properties. Think of shapes made of an arbitrarily deformable substance that can be neither cut nor soldered; the properties are those that are preserved when one arbitrarily deforms a shape so constructed.

In 1858 the mathematician and astronomer August Ferdinand Moebius (1790–1868) described for the first time a new surface in three-dimensional space in a work presented to the Academy of Sciences in Paris, a surface now known as the Moebius strip. In his work, Moebius had described how it is possible to easily construct the surface that now bears his name: take a rectangular strip of sufficiently long paper. If A, B, C, D indicate the vertices of the rectangle of paper, proceed as follows: holding one end of the strip still with one hand (AB for example), perform a twist of 180° on CD along the horizontal axis of the strip so as to make A meet D and B meet C. The construction of the Moebius strip is complete! The surface has been deformed, without cuts or tears, and the rotation performed on one end of the strip has profoundly altered its properties. One consists of the fact that if you trace along the longest axis with a finger, you realise that you run the whole length and return to exactly where you set off, without having to cross the edge of the strip; the Moebius Strip has therefore only one single side, not two, an inside and an outside like a cylindrical surface has, for example. If you wanted to paint the surface of the strip, proceeding along the horizontal axis, it is possible to colour the whole surface without ever lifting the brush and without crossing the egde of the surface. To perform the same operation with a cylindrical surface, you could start with the outside face, but to paint the inside one too you'd have to

Fig. 5. A.F. Möbius, *Zur Theorie der Plyeder und der Elementarverwandtschaft,* in Gesammelte Werke, vol 2, Leipzig 1886, p. 515

cross one of the two edges that separate the outside face from the inside face, something you don't have to do for the Moebius strip. While for the cylindrical surface, if you run a finger along the upper edge you never reach the lower edge, in the case of the Moebius strip if you start from any point on the egde you follow the whole edge and come back to where you started, so therefor it has only one edge. All this has important repercussions from a topological point of view: among other things, the Moebius strip is the first example of a surface on which it is impossible to define an orientation.

First, the newness of the methods used in this new field gave no way to mathematicians to present their results in the traditional deductive form of elementary geometry. Instead, pioneers like Poincaré were forced to base themselves largely on geometric intution. Even today (Courant and Robbins' book was from 1941) a scholar of topology would find that insisting too much on the formal rigors of exposition would lead to easily losing sight of the essential geometric content of a quantity of formal particulars.

The key word is geometrical intuition. Obviously mathematicians have in the course of years attempted to bring Topology into the realm of more rigorous mathematics, but that aspect of intuition remains. It is exactly these two aspects, that of deformations that preserve some properties of the geometric shape, and that of intuition, that have played a deep role in the idea of space and shape, from the 19th Century right up to present day. Several Topological ideas were to be understood by artists and architects through the course of the decades, first by the artists and then much later by architects. It is worthwhile telling the story of the discovery of a topological shape by a great 20th Century artist. A shape that, when the artist discovered it, already existed in the world of mathematical ideas. We're talking about the great 20th Century artist and architect, Max Bill, who died in 1994. Bill tells the following story in the article *How I Began to Make Single-Sided Surfaces* in which he happens to discover the Moebius surfaces (Bill called his sculptures made of Moebius Strips *Endless Ribbons*): [5] *"Marcel Breuer, my old*

Fig. 6. Bill, Max
"Endless Ribbon", diorite
© VG Bild-Kunst,
Bonn 2004

friend from Bauhaus, is the one who is really responsible for my sculptures with one side. Here is how it happened: it was in 1935 in Zurich where, together with Emil and Alfred Roth, he was building the houses of Doldertal, which were very popular in those days. One day Marcel told me that he had been commissioned to build a model of a house, for a show in London, in which everything, even the fire, had to be electrical. It was clear to him that an electrical fire that glows but has no fire is not the most attractive of things. Marcel asked me if I would like to create a sculpture to put onto of it. I started looking for a solution, a sculpture that you could add to the top of a fire and that would hopefully rotate in the rising current of air and, thanks to its shape and movement, would act as a substitute for frames. Art instead of fire! After much experimentation, I found what seemed to me to be a reasonable solution."

The interesting thing to note is that Bill thought he had found a completely new shape. What is even stranger is that he found (invented?) it by playing with a strip of paper, in exactly the same way that Moebius discovered it many years before! *"It wasn't long before someone congratulated me on my fresh and original interpretation of the Egyptian symbol for infinity and the Moebius Strip. I had never heard of either of these things. My mathematical knowledge had never gone beyond the everyday architectural calculations and I didn't have a great interest in mathematics."* Endless Ribbon went on presentation for the first time at the Milan Triennial of 1936. Bill wrote *"Already by the end of the '40s I was thinking of topological problems. From that I developed a sort of logic of shape. The reasons why I was continually attracted to this topic are twofold: 1) the idea of an infinite surface -which is nevertheless finite -the idea of an infinite finite; 2) the possibility of developing surfaces that, as a consequence of the intrinsic underlying laws, lead me almost inevitably to formations that prove the existence of aesthetic reality. But it's 1) and 2) that also indicate another direction. If non-oriented topological structures exist only by virtue of their aesthetic reality, then notwithstanding their exactness, I could never have been satisfied with them. I am convinced that the basis of their effectiveness lies in part in their symbolic value. These are models for reflection and contemplation."*

One could say that just as in four dimensions the object that has sparked the imagination most is the hypercube, or four-dimensional cube, so in the case of Topology the Moebius Strip has taken this role. These shapes that interested Max Bill greatly in the thirties could not fail to interest architects, even if it would take a few years; we had to wait for the widespread use of computer graphics, which allow us to visualise the mathematical objects we're talking about and which support our intuition that otherwise, for nonmathematicians, has trouble manipulating them.

This is what Alicia Imperiale writes in the chapter *Topological Surfaces*: "Architects Ben van Berkel and Caroline Bos of UN Studio discuss the impact on architecture of new scientific discoveries (where 'new' should be considered with a certain indulgence!). Scientific discoveries have radically changed the definition of the term 'Space', attributing a topological form to it. Instead of a static model of constituent elements, space is perceived as something malleable, changeable, and its organisation, its division, its appropriation becomes elastic."

Here is thr role of topology as seen by ab architect: "Topology is the study of the behaviour of a structure of surfaces that undergo deformation. The surface records the changes of the differential space-time shifts in a continuous deformation. This brings further potentialities for architectural deformation. The continuous deformation of a surface can lead to the intersection of external and internal planes in a continuous morphological change, just as in the Moebius Strip. Architects use this topological form in building design by inserting differential fields of space and time into an otherwise static structure."

Naturally, some words and ideas are also deformed when passing from a strictly scientific plane to an artistic and architectural one, seen from a different point of view. But this is not in fact a problem and needn't be a criticism. It is the ideas that circulate freely, and everyone interprets them in their own way, reaping the essence from them, as does Topology. Computer graphics play an indispensable role in all this, allowing us to insert that variable of time deformation that would otherwise be not only unmanageable but unthinkable too.

On the topic of the Moebius Strip, Imperiale continues: "Van Berkel's house, inspired by the Moebius Strip (Moebius House) is conceived as a programmatically continuous structure, that encompasses the continuous mutation of pairs of sliding dialectics that flow one into the other, from inside to outside, from work activities to those of leisure, from load-bearing structure to non-load-bearing structure."

In fact, the Klein Bottle, writes Van Berkel, "can be translated into a channeling system that incorporates all the elements it encounters and makes them precipitate into a new type of internally connected integral organisation." Note the words "integral" and "internally connected" have a precise meaning in mathematics. But this is not a problem here because "the diagrams of these topological surfaces are not used in architecture in a rigorously mathematical manner but rather constitute abstract diagrams, three-dimensional models that allow architects to incorporate differentiated ideas of space and time into architecture."

Max Bill wrote analogous things in 1949 regarding the links between art, form and mathematics: "By mathematical approach we must not mean what is generally called calculated art. Up till now all artistic manifestations have been founded, to a greater or lesser extent, on geometric structures and divisions." Even in modern art, artists have used regulating methods based on calculations since these elements, alongside more personal and emotional elements, gave *balance* and *harmony* to sculpted works. But according to Bill, these methods had however become more and more superficial, since, aside from the exception of the theory of perspective, the repertory of methods used by artists stopped at the Ancient Egyptian era. The new fact occurred at the beginning of the 20th century: "The starting point for a new concept could probably be attributed to Kandinsky who, in his 1912 book *Über das Geistige in der Kunst*, set the premises for an art which would substitute mathematical concepts for the imagination of the artist."

Mondrian, who distanced himself further than anyone else from the traditional concept of art, wrote: "Neoplasticism has its roots in cubism. It could also be called abstract-real paiting because the abstract can be expressed by a plastic reality in painting. This composition of colored rectangular planes expresses the deeper reality that comes through the plastic expression of relationships and not

Fotografie: Christian Richters, Münster

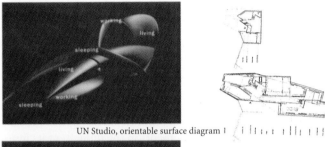

UN Studio, orientable surface diagram

Fig. 7.
Moebius House
by Ben van Berkel
(UN Studio/van
Berkel & Bos),
1993–97

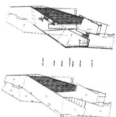

UN Studio, constructive diagram

G. Deleuze, 'The fold in the Soul'

Fotografie: Christian Richters, Münster   Fotografie: Christian Richters, Münster

through natural appearance. [...] The new plastic poses its problems in esthetic balance and thus expresses the new harmony."

Bill believed that Mondrian had exhausted the final possibilities that remained in painting. "I believe it is possible to develop an art based largely on a mathematical concept. [...] Mathematics is not only one of the essential means of primary thought, and therefor one of the recourses necessary for awareness of surrounding reality, but also in its fundamental elements, a science of proportions, of behaviour from object to object, from group to group, from movement to movement. And since this science carries these fundamental elements in itself and places them in significant relationships, it is natural that similar facts can be represented, transformed into images."

Furthermore, Bill adds, these mathematical representations, these limited cases in which mathematics is plastically represented, have an irrefutable aesthetic effect. And here is the definition of what a mathematical concept of art must be: "The mathematical concept of art is not mathematics in the strictest sense of the term, and we could also say it would be difficult for this method to use what is understood as exact mathematics. It is rather a configuration of rhythms and relationships, of laws that have their individual origin in the same way mathematics has its original innovative elements in the thought of its innovators."

To be convincing, Bill needs to supply examples, examples interesting from his point of view as an artist, in other words examples of those that call the mysteries of mathematical problems "the 'indescribable' of space, the distancing and nearing of the infinite, the surprise of a space that begins in one place and ends in another, that is simultaneously the same, limiting with no exact limits, parallel lines that intersect and infinity that returns to itself." The Moebius Strip, clearly.

As we mentioned, architects have, albeit a little belatedly, learned of scientific discoveries in the field of Topology. Besides starting to plan and build, they have started to reflect.

In 1999 in his doctoral thesis *Architettura e Topologia: per una teoria spaziale della architettura* (Architecture and Topology: Through a Spatial Theory Of Architecture) Giuseppa Di Cristina writes: "The final conquest of architecture is space: this is generated through a sort of positional logic of the elements, i.e. through an arrangement that generates spatial relations; the formal value is thus substituted by the spatial value in the configuration: what is important is not so much the aspect of the exterior form as its spatial quality and therefore topological geometry of non-rigid figures with no 'measurements'. This is not something purely abstract that comes before architecture, but rather the tracks left by that modality of action in the spatial concretisation of architecture."

A volume of articles was published in 2001 on the theme "Architecture and Science" [6]. In the preface *The Topological Tendency in Architecture* by Di Cristina, it is explained that "The articles here bear witness to the interweaving of this architectural neo-avant-garde with scientific mathematical thought, in particular topological thought: although no proper theory of topological architecture has yet been formulated, one could nevertheless speak of a topological tendency in architects at both theoretical and operative levels. [...] In particular developments in modern geometry or mathematics, perceptual psychology and computer graphics have influenced the present renewal of architecture and the evolution of architectural thought. What most interests architects who theorise about the logic of curvilinearity and pliancy is the meaning of 'event', 'evolution' and 'process', that is, of the dynamism that is innate in the fluid and flexible configurations of what is now called 'topological architecture'. Architectural topology means the dynamic variation of form facilitated by computer-based technologies, computer-assisted design and animation software. The topologising of architectural form according to dynamic and complex configurations leads architectural design to a renewed and often spectacular plasticity, in the wake of the Baroque and of organic Expressionism."

Here is what Stephen Perrella means by "architectural topology": "Architectural topology is the mutation of form, structure, context and programme into interwoven patterns and complex dynamics. Over the past several years, a design sensibility has unfolded whereby architectural surfaces and the topologising of form are systematically explored and infolded into various architectural programmes. Influences by the inherent temporalities of animation software, augmented reality, computer-aided manufacture and informatics in general, topological 'space' differs from Cartesian space in that it imbricates temporal events within form. Space then, is no longer a vacuum within which subjects and objects are contained, space is instead transformed into an interconnected, dense web of particularities and singularities better understood as 'substance' or 'filled space'. This nexus also entails more specifically the pervasive deployment of teletechnology within praxis, leading to a usurping of the real (material) and an unintentional dependency on simulation." Ideas on geometry, Topology, computer graphics and space-time come together in these observations. The cultural links have, in the course of the years, worked: new words, new meanings, new associations.

Final Observations

I have sought to talk of several important moments that have led to a mutation of our understanding of perception of space, seeking to gather not only the technical and formal aspects that are also essential in mathematics, but also the cultural aspect, by talking of the idea of space in relation to some aspects of contemporary architecture. I would like to recall only two words that are of great importance: fantasy and liberty. It is perhaps these two magical words that have allowed contemporary architecture to greatly enrich design heritage. Fantasy and liberty that come from the coming together over the course of years of many elements: computer logic, new geometries, Topology and computer graphics. For, even if very few people realise it, mathematics is, or can be, the kingdom of liberty and fantasy.

Without all this, the planning of the *Museum of The Hellenic World* would have been unthinkable. A culture started in that place thousands of years ago and that would be celebrated in that same place with a construction highly symbolic of the history of the culture of the Mediterranean.

References

[1] Imperiale, A.: *New Bidimensionality,* Birkhaüser, Basel (2001)
[2] Hodges, A.: *Storia di un Enigma,* Bollati Boringhieri, Turin (1991)
[3] Emmer, M.: *Mathland: from Flatland to Hypersurfaces,* Birkhaüser, Boston (2004)
[4] Courant, R. and Robbins, H.: *Che cosa è la matematica?* Bollati Boringhieri, Turin (1971)
[5] Bill, M.: *A Mathematical Approach to Art,* (1949) reprinted with corrections from the author in M. Emmer, ed., *The Visual Mind: Art and Mathematics,* Boston, MIT Press (1993) Quintavalle, A., ed. *Max Bill,* book n. 38, Dipartimento Arte Contemporanea, Università di Parma (1987)
[6] Di Cristina, G., ed., *Architecture and Science,* Wiley-Academy, Chichester (2001)

On the links between mathematics and culture:
− M. Emmer, ed., *Matematica e Cultura 2000,* Springer verlag Italia, Milano, (2000); English edition, Springer verlag, Berlin (2004)
− M. Emmer, ed., *Matematica e Cultura 2001,* Springer verlag Italia, Milano (2001)
− M. Emmer, ed., *Matematica e Cultura 2002,* Springer verlag Italia, Milano (2002)
− M. Emmer, ed., *Matematica e Cultura 2003,* with musical CD, Springer verlag Italia, Milano (2003); English edition in preparation.
− M. Emmer, M. Manaresi, eds., *Matematica, arte, tecnologia, cinema,* Springer verlag Italia, Milano (2002); 150 pages are dedicated to cinema fiction and mathematics; English edition, updated with films from 2003, Springer verlag, Berlin (2004)
− M. Emmer, ed., *Matematica e Cultura 2004,* Springer verlag Italia, Milano (2004); English edition in preparation.
− M. Emmer, ed., *Matematica e Cultura 2005,* Springer verlag Italia, Milano, in preparation.
− Web *"Matematica e Cultura":* http://www.mat.uniroma1.it/venezia2005 (the date in October changes each year)
− M. Kline, *Mathematics in Western Culture,* Oxford University Press, New York (1953)

On the links between mathematics and achitecture:
− M. Emmer, *Mathland, from Flatland to Hypersurfaces,* Birkhaüser, Boston (2004)
− *Topologia e Morfogenesi,* ED La Biennale, Venezia (1978)
− Ben van Berkel, *Mobile Forces / Mobile Kräfte,* Ernst & Sohn Verlag, Berlin (1994)
− John Beckmann, ed., *The Virtual Dimension: Architecture, Representation, and Crash Culture,* Princeton Architectural Press, New York (1998)
− G. Di Cristina, ed., *Architecture and Science,* Wiley Academy, Chichester (2001)

On the links between mathematics and art:
− M. Emmer, ed., *The Visual Mind: Art and Mathematics,* The MIT Press, Boston (1993)
− M. Emmer, *La perfezione visibile,* Theoria, Roma (1991)
− M. Emmer, ed., *The Visual Mind 2: Art and Mathematics,* The MIT Press, Boston, to appear (2005)
− M. Emmer, *Art and mathematics,* 20 videos realised with artists from all over the world http://www.mat.uniroma1.it/people/emmer .

Visual Mathematics and Computer Graphics

Introduction:
Visual Mathematics and Computer Graphics

Michele Emmer

In the seventies a ingeneer, Frank J. Malina, former collaborator of Von Braun in the creation of ballistic missiles decided to become a kinetic artist and move from USA to Paris. He was very interested in the relationships between technology, science and the arts, as much because of his artistic training. At that time he was to found the journal *Leonardo,* which would quickly become the most important international journal on the relationships between science, technology and art. For some years now the journal has been edited by Frank Malina's son, Roger Malina, an astrophysicist. It is published by MIT Press. It was Frank Malina who was very drawn towards mathematics. "In one way or another, mathematics underlies the ideas and artwork discussed in this volume – only elementary echoes are evident in some, but there are sophisticated applications of the *queen of science* in others. To understand the non-figurative or abstract artworks presented requires, in general, a comprehension of messages of a kind different from those that artists have traditionally used. It would be foolhardy to predict at present that such understanding will cause many art lovers to obtain aesthetic satisfaction from them.

What does seem certain is that theorising about the visual arts will be transformed by procedures involving, in particular, cybernetic conceptions and that prodigious machine called a digital computer."

These words were written by Frank J. Malina in June 1977 in Boulogne sur Seine near Paris in the preface to the volume *Visual Art, Mathematics, & Computers.* [1]

In the introduction to the volume, Frank Malina wrote:

"Can one look for a synthesis of art and mathematics? Prior to the emergence in the 20th century of certain developments in science and technology, in particular mass communications systems, cybernetics and electronic digital computers, such a synthesis may have been limited to an amplification of what a few artists had done in the past. But this would not have amounted to a synthesis as normally understood. Will these new developments bring about a synthesis of art and mathematics?

Digital computers offer many possibilities for the production of both figurative and non figurative 2-and 3-dimensional visual art. Some artists welcome the computer not only as a means for quantity production of objects of art but as an aid to creativity. However, at this stage, it seems to me that many are fascinated more by computer technology than by mathematically oriented visual art, exceptions will be found in this book. Aesthetics is seen in a different light when viewed from the points of view of cybernetics, information theory and computer programming, which involve the application of numerous branches of mathematics. It remains to be seen whether artists, in general, will take a more serious interest in *cybernetic* aesthetics than they did in the *theories* of aesthetics of the past and, thereby, arrive at a synthesis of art and mathematics".

Words written in 1977.

A few years later one of the author of the book edited by Frank Malina (and one of the author of *The Visual Mind* I have edited in 1993 [2] and author of a paper included in this volume) Herbert W. Franke, wrote as an invited contribution to the volume *The Beauty of Fractals*: [3]

"Art critics in the centuries to come will, I expect, look back to our age and come to conclusions quite different that our own experts. Most likely the painters and sculptors esteemed today will nearly have been forgotten, and instead the appearance of electronic media will be hailed as the most significant turn in the history of art.

It will be pointed out that it became possible for the first time to create three dimensional pictures of imaginary landscapes and other sciences with photographic precision, and with these pictures not just to capture an instant in time but to include the reality of change and movement.

New technologies of picture processing were developed from older methods of picture analysis and pattern recognition based on methods from photography. These techniques allow images acquired from science, technology and medicine to be more easily evaluated, and in some cases, to be possible for the first time.

Some mathematicians and programmers used aesthetic possibilities of graphics system form the very beginning in the early sixties. Most avoided using the word *art* in connection with their work, however, thus dodging a conflict with the art establishment.

Happily, the representatives of this new direction were little influenced by these discussions and continued their work independently of any theoretical objections. Indeed, an impressive repertoire of computer graphics has accumulated, and this collection reflects the technical progression from simple drum plotter to high-resolution visual display graphics."

References

[1]   F. J. Malina, ed, *Visual Art, Mathematics & Computers,* Selections from the Journal Leonardo, Pergamon Press, Oxford (1979).
[2]   M. Emmer, ed, *The Visual Mind: Art and Mathematics,* The MIT Press, Cambridge, Mass (1993).
[3]   H.-O.Peitgen & P.H. Richter, *The Beauty of Fractals,* Springer verlag, Berlin (1986).

Math Awareness Month 2000:
An Interactive Experience

THOMAS F. BANCHOFF and DAVIDE P. CERVONE

Introduction

Mathematics Awareness Week was founded in 1986 by the Joint Policy Board for Mathematics (JPBM) as a public relations effort to increase appreciation for the power, scope and beauty of mathematics. In 1999, the campaign was extended to Mathematics Awareness Month, held annually in April.

Each year, Mathematics Awareness Month is the responsibility of one of the three organizations that make up the JPBM: the American Mathematical Society (AMS), the Mathematical Association of America (MAA), and the Society of Industrial and Applied Mathematics (SIAM). For MAM2000, that role fell to the MAA.

Math Awareness Month always has a theme approved by the JPBM as a whole. Themes in the past have included "Mathematics and Symmetry", "Mathematics and Finance", and "Mathematics and the Internet". In 1999 and 2000, the first-named author was President of the MAA, and he suggested "Math Spans All Dimensions" as the theme for 2000. This concept was approved in May 1999 and the preparations began.

The centerpiece for each MAM campaign is a poster, designed to be displayed in classrooms and on bulletin boards at colleges and universities, to draw attention to the theme and to lead students and teachers to delve further into the topic by reading ancillary materials, usually in the form of short essays on sheets distributed along with the poster. For MAM2000, we decided to design and implement an electronic poster that would put the additional materials directly at the hands of the viewer, and appeal to audiences at different levels.

The key metaphor for the electronic poster is that of an Advent Calendar. Such a calendar usually is made from cardboard, containing small rectangles numbered for the days of December leading up to Christmas. The rectangles are flaps, and each day one flap is opened to reveal some toy or other present, a promise of things to come.

Our idea was to create an electronic analogue of such a calendar, with small rectangular icons or pictures that could be selected to open up more and more information and further links. Other images and words on the poster would be "live" as well, inviting exploration into other aspects of the theme. From the beginning, we intended to develop a central geometric object that would organize the theme, accompanied by pictures of individuals associated with "Mathematics Across All Dimensions".

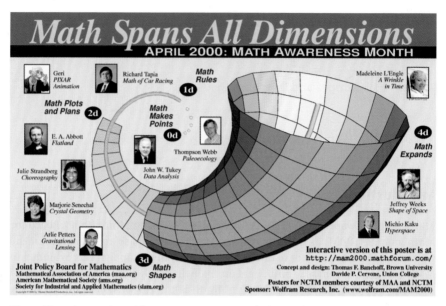

Fig. 1. The print version of the MAM2000 poster includes a reference to the electronic version [11], where each picture and caption is a link to more information

From the outset, the two authors collaborated by phone and email to bring this idea to fruition. The production of the MAM2000 poster involved a number of challenges, both technical and organizational, and the story of its development and implementation is the subject of this paper. In the sections below, we treat both the technical aspects and the ones dealing with the participants.

Beginnings

On one afternoon in the fall of 1999, when the first-named author was visiting the second at Union College for a day, many of the basic design decisions fell into place. First, we settled on the central organizing image, a "cornucopia" formed by expanding circles perpendicular to an exponential spiral, somehow symbolizing the progression from zero to four dimensions. Our initial sketches suggested that the poster should be horizontal rather than the usual vertical arrangement. The practical size for a bulletin board poster was set at 11" by 17", so ours was 17" by 11". For mailing purposes, the poster would be folded so it could fit in a standard envelope. We wanted to avoid having the crease intersect the central cone in an inconvenient way, and that determined the image placement in our design. Finally, we sketched out the arrangement of pictures around the cone, as well as the placement of acknowledgements and sponsorships. Remarkably enough, that initial basic sketch survived all the refinements and changes that we made over the next six months.

The selection of spokespersons whose picture would appear on the poster raised some challenges. We wanted to be unapologetically representative in our choices, with respect to fields and applications, and also with respect to age, gender, and race. We had a good pool to choose from in all categories.

Paul Tukey is the father of interactive multidimensional analysis. He had participated actively in our 1984 interdisciplinary symposium, "From Flatland to Hypergraphics". He was happy to be included, but he felt that he would not be able to contribute much that was new. We already had material from him for the earlier symposium and we were counting on help from other researchers in exploratory data analysis. That set up the first picture in the 0-dimensional section, subtitled "Math Makes Points".

Thompson Webb III is a colleague of the first author in the Geology Department at Brown University who has participated in a variety of interdisciplinary activities related to space and time. He contributed a major section on core-sample research in paleoecology to the author's volume *Beyond the Third Dimension* [2]. This section maps the prairie-forest boundary within the midwest United States over a ten-thousand-year period by analyzing fossilized pollen concentrations in yearly depositions in lake beds. Since that chapter and its associated materials were already prepared, this was an obvious choice for the beginning of the 1-dimensional section, tentatively titled "Math Rules".

Geri, the star of the Oscar-winning short animated film "Geri's Game" [9], also making a guest appearance in the feature film "Toy Story II" [10], is the only fictional person pictured on the electronic poster. Tony de Rose and Ed Catmull of PIXAR Studios met the author at a conference on computeraided geometric design in November 1999 and both agreed to participate in the project. They arranged to have Geri featured on the poster, with links to articles on the use of mathematics in computer animation, in particular the frontier research topic of subdivision surfaces. It was a definite plus that some of the images in the article about subdivision surfaces related directly to the Ph. D. thesis problems of both authors. This was the first topic in the 2-dimensional section, tentatively titled "Math Plots and Plans".

Edwin Abbott Abbott is the only deceased person who was chosen to appear in the gallery. As the author of *Flatland* [1], he is the founder of the dimensional analogy for understanding and communicating between dimensions. In addition, just the previous year we had assembled materials for a Brown University Library exhibit, "*Flatland*, a Millennial Book" in conjunction with the MAA 1999 MathFest held in Providence, RI. We already were committed to producing interactive Internet-based materials connected with the publication history of Flatland and the life and times of its remarkable author. We previously had accumulated a great number of links about this topic, in particular electronic records of final projects by students in the first author's frequent courses on "The Fourth Dimension".

Julie Strandberg, chief choreographer at Brown University, is another longtime collaborator of the first author. An account of her dance composition "Dimensions", a retelling of Flatland themes, already appeared in a section of *Beyond the Third Dimension* [2]. There were other links available to sites on modern dance, and symbolic notation for classical and modern dances in a variety of cultures.

Marjorie Senechal is a world leader in mathematical crystallography and an organizer of a major conference on polyhedra. She agreed to write about her recent work and to provide photographs of a collaboration with the mathematician and sculptor Helaman Ferguson. This last topic, with its connections both to planar tilings and to solid crystals, is the bridge to the 3-dimensional part of the poster, with the title "Math Shapes".

Arlie Petters, a mathematician specializing in astronomy, specifically in the theory of gravitational lensing, has interacted with the first author for a number of years, and he had produced excellent visual presentations of his work. Despite a busy schedule, he was quite willing to participate and to help us find links to other related sites.

Michio Kaku, author of the recent volume *Hyperspace* [6] and numerous other writings on modern mathematical physics, had contacted the first author concerning background images for a television interview. He was also happy to participate and suggest a number of links to other sites, although he did not have time to develop new materials for the poster.

Jeff Weeks could be counted on to contribute at all levels since he already had produced impressive and accessible materials on topology and geometry, especially in three and four dimensions. He had just been named a MacArthur Fellow, and he was willing not only to appear on the poster, but also to help with the writing and the production of the interactive applets for viewing 4-dimensional phenomena. His section fit in with 4-dimensional geometry, with the title "Math Expands".

Madeleine L'Engle is a colleague with whom the first author has worked for years, primarily in connection with her extremely well-known prize-winning book, *A Wrinkle in Time* [7]. For many students, this fantasy was their first introduction to the fourth and higher dimensions and to the 4-dimensional cube, also known as the *tesseract*. One of her granddaughters, who is currently helping her with her projects, was very encouraging and helpful in arranging for her participation in the poster. There were many links available for this site, including a final project by a group of students in a course the first author taught at Yale as a visiting professor in the fall of 1998.

All of the contacts with those who appeared on the poster were made in the mid-fall of 1999. Aside from their connection with dimensionality in their work, the participants have a number of other things in common. First, the first author had worked with all of them over the years and had maintained good contact with them. This was essential considering the short amount of time we had for assembling the list of participants. Second, most reported that they had little or no time to work on producing anything new for the project, so it was clear that we ourselves would be doing most of the writing and producing demonstrations.

Only one person who had been contacted declined to participate, since the work of that person over the past ten years had veered away from topics connected with the theme of the poster. That meant that there was one slot left to be filled. Several colleagues who were consulted suggested the same person for a position in the 1-dimensional part of the poster, namely an applied mathematics and engineering professor who was also known as a champion drag racer! Richard Tapia was recommended as well for his exemplary work for access to higher

Fig. 2. The page about Jeff Weeks describes him and his work, and includes links to software that he produced, articles in which he is featured, his book *The Shape of Space* [8], and movies related to that book.

education for Latino and Native American students, and he saw this poster as fitting in with his work on several levels. His assistant provided us with excellent materials for the poster.

Our selection process for the persons to appear on the poster was now complete, with a few months still to go.

The next task in the project was to make arrangements with the MathForum [12], which had provided the web server for earlier Math Awareness Month (MAM) materials. This became a straightforward matter, without complications, due to the excellent work of Gene Klotz, with whom we had worked on other projects over the years. Although our concept of an interactive electronic poster was orders of magnitude more involved than previous MAM efforts, we received full cooperation from the MathForum leaders and technicians in setting up MAM2000.

In recent years, there has been corporate sponsorship for the MAM poster, and this year negotiations were already underway with Wolfram Research, Inc. The basic concept that we proposed found favor, and we worked with Paul Wellin for the next several months on various aspects of the project. One that took more time than we expected had to do with the central image. We had produced our prototype using the *CenterStage* program developed by the second-named author, but it was considered appropriate to render the central figure using Mathematica. One of the designers at Wolfram also proposed a central image along with a color scheme and background design. These were brought, along with the sketch by the authors, for consideration by the sponsoring Joint Policy Board for Mathematics. That body accepted the color and background proposed by the designers from Wolfram, but they preferred the simpler central figure of the authors as fitting in better with the theme of proceeding through various dimensions.

87

Technical Development

Once the nature of the central figure was determined, we proceeded to work on several levels simultaneously. We finalized the mathematical design of the exponential cone and worked on the layout and organization of the poster itself. We began accumulating photos of the participants, and we produced drafts of the pages describing the people and their work. The majority of the pages were written by the first-named author, assisted at various times by Jeff Weeks and Marjorie Senechal. As the pages took shape, we began the search for appropriate links to add to these pages. Several students in the first author's class helped to identify these links, and Frank Farris not only located materials but also helped to design a page of resources on two-dimensional patterns. The second author handled nearly all of the technical aspects of the site design and implementation, organizing the layout of all the pages, and producing the movie clips that are part of the site. Most of the interactive applets were produced by students at Brown University, while one applet for manipulating the hypercube was developed by Jeff Weeks specifically for this project.

To indicate the progression of dimensions along the central cornucopia image, we employed a series of different forms along the central exponential curve. We represented zero dimensions by a string of points along the curve, becoming a thickened curve to depict one dimension. For two dimensions, we displayed the curve together with a segment along the principal normal vector (the segment in the plane of the velocity and acceleration vectors of the curve, perpendicular to the velocity vector). The segment was centered on the curve and we chose to have it grow linearly rather than exponentially, for artistic reasons. To represent three dimensions, we used circles centered on the curve in planes perpendicular to the curve, with radius continuing to grow linearly. We used an exponential spiral with a carefully chosen exponent, and a linear function determining the radii of circles so that this conical surface would open just the right amount for the space available on the poster. Although it is not immediately apparent, the exponential spiral is not a curve in the plane but rather a space curve, where the third coordinate is a linear function of the parameter that defines the curve. This helps create the effect that the cone is "opening up" at the end, so that we can see into it. The viewer is encouraged to imagine how the cornucopia would continue into the fourth dimension and higher!

In the final electronic version of the poster, the formulas we used are included in one of the links that appear when any point on the central figure is selected. In all of these pages, we attempted to describe the objects in ordinary language initially, but provided the more sophisticated reader with the opportunity to obtain progressively more technical discussions of the material. We reproduce the key paragraphs of that analysis here [13]:

> To describe our central image in technical language, we have to define a central curve $X(t)$ over a parameter t. The curve we chose is an arithmetic spiral lifted onto a cone, with equation
>
> $$X(t) = (t\cos(t), t\sin(t), \pi t).$$

The radius function $r(t)$ that describes the expanding cone around this spiral is given by $r(t) = 0.8e^{0.6t}$.

As t goes from 0 to $\pi/2$, we show a number of points on the curve. From $\pi/2$ to π, we show the curve itself, actually represented by a thin tube.

From π to $3\pi/2$, we show a "normal strip" of the form $X(t)+rP(t)$ where r goes from $-r(t)$ to $r(t)$. Here $P(t)$ corresponds to the principal normal, a unit vector perpendicular to $T(t)$, the unit tangent vector of the curve, and lying in the plane determined by the velocity and the acceleration of the curve.

From $3\pi/2$ to 2π, the figure is an expanding cone, or "cornucopia". The equation of the circular slice for each t is

$$X(t) + r(t)(\cos(u)P(t) + \sin(u)B(t))$$

where $B(t)$ is the unit binormal vector perpendicular to $T(t)$ and $P(t)$, and where u is a parameter that goes from 0 to 2π.

Thus the image grows not only in size, but also in dimension.

As we finalized our design for the central cone, we began to put the images of the participants in place on the poster. This was an evolving process. New pictures were added as we found them and obtained permission to include them. We produced a preliminary poster using low-resolution images from the web as a sample to show to the JPBM at their December 1999 meeting. After the design was approved, the MAA selected the poster as the cover art for the first full-color issue of their news publication *Focus* [4] in March of 2000. The choice was made just a few days before the final artwork was needed by the publishers, so we had to move quickly to get everything together in time for their deadline. In particular, we did not at this point have any of the high-resolution versions of the images of the participants. Remarkably, we were able to obtain photos or scans from everyone within just two days of our request! This allowed us to put the final print version together in time for use by *Focus*. Their original design called for the poster to be used for both the front and back covers (the 17 × 11 image would be folded in half to form both covers); but the large end of the cone made for a less successful image than we had anticipated, and the editors decided to put the entire poster on the front cover. The reduced size would have made the images of the participants too small to reproduce well, so we chose to print only the central cone together with the wording from the poster. The result was extremely successful, and generated numerous requests from MAA chapters and other organizations to use the image on their own publications.

Final touches were made to the poster during the following week, and by mid February the main poster was ready to go to print. An initial run of 4,000 posters was produced and mailed to mathematics departments in colleges and universities around the country. A second run of 120,000 was sponsored by the MAA and the National Council for Teachers of Mathematics (NCTM) and mailed out to all NCTM members during the first week of April.

89

Putting It All Together

Once the print version of the poster was complete and at press, we were left with a month and a half to put the electronic version together before the April 1st opening date. In reality, we needed to be on line much earlier. All parties concerned were anxious to have a version available before the beginning of April, especially since the site had been announced in the March issue of Focus [4]. The need to make as much of the site "live" as possible, while still allowing for rapid change and prototyping kept both authors very busy (while they were performing normal teaching duties in addition to developing this site).

Our vision for the design of the interactive web site was based on several themes. First, we wanted a multi-level approach, through which any reader could have access to general information about the topics of interest, and could continue to follow links that would bring them more deeply into the subject. Thus, for example, someone interested in the main cornucopia image could click on that to get a general description of the object, and then could follow additional links to the details of the equations involved, or to movies and interactive versions of the cone. A person interested in dance could see photographs of a dance sequence, or sites devoted to dance notation or ballet.

Second, since the site was to be accessed by students in classrooms around the country, we wanted the pages to load quickly and be viewable by a wide variety of browsers. This meant that we wanted to use a minimum of "cutting-edge" HTML features, to make the material accessible to the widest possible audience, while still forming an attractive site. Aside from the opening page with its large image of the poster, most pages used only small images that download relatively quickly. Furthermore, large or timeconsuming items, such as movies and Java applets, do not appear unless the reader specifically requests them. For example, the movies of the central cone appear on a separate page specifically for these movies, not on the one linked to the cone from the main poster. This approach also made it possible to provide the movies in several formats, so the reader can select the one that is most appropriate for his or her connection and software configuration.

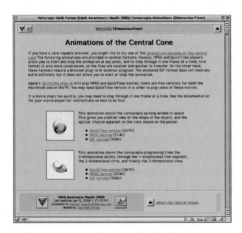

Fig. 3. The page of movies for the central cornucopia provides these movies in several formats, and describes what player is needed for each

Third, we wanted to be sure that there was a consistent "look and feel" to the site. In particular, we wanted all the pages to have connections back to the main poster page and the MathForum. Since many users would be arriving at the site by following links from search engines, it was crucial to give them a clear idea of where they had arrived, and how the page they found fit into the structure of the MAM2000 site. We wanted to avoid the frustration of landing on a page that does not provide any connections to the rest of the material at that site.

We paid particular attention to the navigation buttons for our site. The header was designed to make it easy to move from page to page or back to the home page. These buttons were placed at the top of the page, so that they could be used without the need to scroll the page, and they are small and unlabeled so they don't take up much room. At the bottom we placed similar buttons, this time including the titles of the pages to which they refer. They are at the bottom, so that someone who has read through the document would not need to scroll back up to the top to get at them. We included the titles at the bottom of the page as well, to give more information about possible continuations.

Fourth, we wanted to use the structure of the site to mirror the structure of the material being discussed. The main theme, that of the sequence of dimensions, appears in several places within the poster. We have already seen how the cornucopia displays this progression. Originally, we imagined that there might be a sequence of images on the poster itself, going from a point to a segment to a square to a cube and then to a hypercube, perhaps sitting inside the cornucopia. This idea was abandoned since it was felt that the design was already quite busy, and it became clear that it was possible to develop this and other notions in a series of small essays directly accessible from the labels "0d", "1d", and so forth. Each of these short essays gave an overview of the images associated primarily with that dimension, and introduced some further analogies. In each case there was an attempt to distinguish between the "intrinsic" dimension of an object and the "extrinsic" dimension of the "ambient space" which contained it. The intrinsic dimension of an object is the number of parameters necessary to locate each point along the object, while the "extrinsic" dimension of an object in space is determined by the number of coordinates in the space where the object is situated.

In addition to being linked to the dimension labels on the main poster, each of these essays was linked to the next. In a similar way, we linked sequentially the pages about the various people featured on the poster, so readers could continue to the next person without having to return each time to the poster itself.

One more sequence of this sort was developed within the dimensional essays connected with the "dimensional buttons" labeled "0d" through "4d" on the poster. These include a sequence of interactive demonstrations on rotations of cubes and hypercubes. In dimension two, a square with four vertices $(\pm 1, \pm 1)$ can be rotated while a matrix shows the coordinates of the images of the first and the second basis vectors. A similar demonstration in dimension three represents a cube with its eight corners at $(\pm 1, \pm 1, \pm 1)$, and allows the viewer to rotate the cube interactively and see the rotation matrix involved in producing that view. Finally, a demonstration for the hypercube extends these ideas into the fourth dimension.

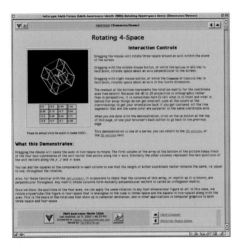

Fig. 4. This Java applet demonstrates rotations in four dimensions; the matrix at the bottom shows the current rotation. The instructions for the user are to the right of the applet, and a mathematical discussion follows. This page is linked to the applets for the other dimensions, as well as to the essay on the fourth dimension

Since these demonstrations are Java applets, they appear on pages of their own, complete with instructions on how to use them, and are linked from the essay on the appropriate dimension; but they also are linked to each other so that a user can go easily from experimenting with one dimension to investigating the analogous phenomena in the next.

Another important portion of this site is not linked directly to the poster itself, but appears in the list of links below it. This is the collection of "essays on dimension". Two chapters from the first author's book, *Beyond the third dimension* [2], are reproduced here. We were able to obtain an electronic copy of the text from the publisher (with their permission), and since the second-named author had produced the line art for the book, we were able to regenerate these images for the web fairly readily. Several pictures required more work, however; these were originally produced using programs that were no longer available to us, and so they had to be reproduced in more modern software. For example, the three images

Fig. 5. Several chapters from publications about dimensions are reproduced here, for viewers who are interested in more extensive details

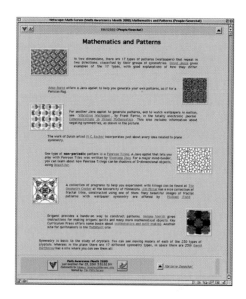

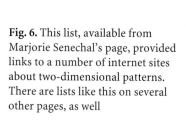

Fig. 6. This list, available from Marjorie Senechal's page, provided links to a number of internet sites about two-dimensional patterns. There are lists like this on several other pages, as well

showing the swallowtail catastrophe surface are not the original ones from the book, but were recreated specifically for the MAM2000 site.

Similarly, a chapter from *On the Shoulders of Giants* [3], published by the Mathematical Sciences Education Board of the National Research Council, also is included in this collection. Again, the text and graphics had been produced by the authors, so we were able to create the pages and images for the site quickly. The last essay, A. K. Dewdney's "Four-Dimensional Dementia" [5], originally appearing in *Scientific American* in 1986, required us to scan the artwork, producing somewhat lower-quality results.

These essays provide more extensive background for those interested in pursuing the idea of dimension further. Other resources are included in the form of lists of videotapes about dimensions, titles of books about dimensions, and links to web sites related to dimension.

The Final Touches

As April approached, the web site began to receive more attention from outside readers. To make it easier for other sites to link to the MAM2000 poster, we provided MAM2000 icons in several sizes, together with HTML code that could be copied directly into a web page to produce a link to the MAM site. These were used by dozens of math departments in conjunction with their Math Awareness Month activities, and the images also appeared on other math-related websites throughout the country.

One of the original issues we faced when we began to develop the MAM2000 site was how large to make the main image. We wanted the poster to be able to fit on the small-sized screens at schools with fairly old equipment (e.g., screens that

93

Fig. 7. This image was made available for web administrators to include on their mathematics pages as a link to the MAM2000 site. We even provided the HTML code to make the links

are 640 × 480 pixels). On the other hand, we wanted to provide a larger image for those with access to larger screens. We ended up providing three separate solutions to this problem.

First of all, we produced the main image in three different sizes, 612 × 396, 765 × 495 and 918 × 594. The viewer could switch between these sizes by clicking on one of the "poster size" buttons to the right of the image of the poster. Thus if the user wanted to use a larger (or smaller) version, it was possible to select the most appropriate one for the equipment at hand. The pages were set up so that going back to the home page would bring the reader back to the size last selected.

Second, we arranged that when the MAM2000 home page was first loaded, it would run a JavaScript program that would attempt to determine the size of the browser's window and select the largest of the three sizes that would fit comfortably on screen. This works properly for Netscape and Internet Explorer, provided JavaScript is enabled; the correct page would be selected automatically without the user even being aware that a choice was being made. For those without a browser that processes JavaScript (or who have it turned off), the 612 × 396 version would be used by default.

We gave visitors with a small screen, who would like to see an enlarged version of the poster, a third mechanism to make this possible. To the right of the poster are a set of small icons representing four quadrants of the poster. Clicking on a quadrant enlarges that region of the poster by a factor of two in each direction. All the links to portions of the poster that appear on a magnified section still work, and the viewer can move directly from one quadrant to another without returning to the full poster.

Fig. 8. Visitors to the MAM2000 site can choose to enlarge one quadrant of the poster. The links from the poster are all still available, and any of the other three quadrants can be selected via the buttons at the right

While it is somewhat risky to make web pages too "smart", this scheme seems to have worked out well. The browser makes an educated guess about what size and view the user wants, and then that size can be modified as desired. Those with older browsers (which do not process JavaScript) can still have access to the larger sizes without even knowing they are missing anything.

With these final changes in the last week of March, we were ready to "go live" on the 1st of April, as planned.

The Site in Use

The opening of the MAM2000 site on April 1st was a success, and we received numerous complimentary mail messages that month. We were eager to see the usage statistics collected from the log data from the web server, and were kindly given the raw data by the staff of the MathForum. It is not easy to interpret such information, however. Most people simply discuss "web hits", which is the total number of files shipped by the server, but this is a poor measure of the usage of the site for a variety of reasons. First, each image on a page is shipped separately, so a page with many pictures will generate many hits, artificially inflating the hit count. Similarly, a single Java applet may cause the downloading of numerous .class-files. Second, many web-based search engines send out "robots" to read pages on the web so that they can be indexed and included in the search engine's database. Some of these robots are quite aggressive, reading the entire web site on a regular basis. Such hits should not count as "real" hits, since they don't correspond to people actually reading the web site. Depending on how well linked to search engines a site is, this can be a significant portion of the traffic from the web site. Finally, because browsers and proxy servers can cache commonly loaded pages, the log data may not include all the times that some users actually *do* view the pages.

For all these reasons (and more), any web log data should be viewed with a healthy dose of skepticism, and used only as a rough idea of the actual usage. Filtering the data to attempt to remove the effects can make a big difference in the results. While it is fairly easy to remove images from the hit count, there is no reliable way to distinguish "real" hits from those made by automated robots, and no way at all to count the hits that are serviced by caches and proxy servers.

Rather than a pure hit count, a more interesting statistic would be a "session count" indicating how many people viewed the site over a given time period. Such information is extremely difficult to come by (without special server software). The information in the log says that a file was shipped to a particular machine at a particular time, but one can only make educated guesses about when several such events fit together into one "session" by a single user at that machine. One approach is to use the times that the files were sent to a given machine, together with information (not always available) about the previous page being viewed by the user to try to piece together the sessions. This is somewhat inaccurate, but does give a reasonable sense of the number and length of sessions for the web site.

In our case, the data obtained from the MathForum included hits to other parts of their web site that needed to be removed, and seemed to be missing some of the hits to the MAM2000 pages. (Log data indicated that a link was made from another page of the MAM2000 site, but no record was included of that page being served to the indicated machine.) The following data gives the best information we have about the usage of the MAM2000 pages.

In the month prior to the official opening of the site, we averaged about 800 (non-robot) sessions or about 2,100 (real) pages per week. Not surprisingly, in the first week of April we reached our peak usage of about 3,900 sessions (16,000 files) per week. The following weeks showed a decrease to 3,100, 2,800, and 2,300 sessions per week, and after April, there was a significant drop off (as expected), to about 500 sessions a week during the following month. The total usage in April was approximately 12,200 sessions (52,700 files), or about 400 sessions a day of around 4.3 pages each. The usage by robots climbed steadily in the month of May, until it accounted for about a third of the total usage of the site, with more than 1,500 files being shipped to over 30 different robots per week.

Conclusion

Going beyond the traditional poster format to an electronic document involved a number of challenges to be met over a short period of time. Although the MAM2000 poster is still accessible [11], the number of visitors is down to a very small number. The ephemeral character of the project becomes clear as current visitors encounter more and more links to external sites that simply do not work. We tried hard to keep things operating during the actual Math Awareness Month, and for the rest of the year 2000, but now it seems inevitable that the outside links will decay faster and faster. Thus it is hard to archive this kind of electronic document. To achieve any permanence, we would have to collect all the files to which we linked, a difficult task under any circumstances for a project of this size. It would have been impossible to secure the required permissions to accumulate these files within the time frame for the project. We hope sometime in the future to explore creating a permanent, self-contained version of such a two-dimensional window on "Math Across All Dimensions", based on our experience with Math Awareness Month April 2000.

References

[1] A. E. Abbott, *Flatland, a Romance of Many Dimensions,* Princeton University Press, Princeton, NJ, 1991.

[2] T. F. Banchoff, *Beyond the Third Dimension: Geometry, Computer Graphics, and Higher Dimensions,* Scientific American Library, W. H. Freeman & Co., New York, 1990.

[3] T. F. Banchoff, "Dimension", in *On the Shoulders of Giants,* ed. Lynn Arthur Steen, MSEB National Research Council, National Academy Press, Washington, D.C., 1990.

[4] T. F. Banchoff and D. P. Cervone, "Cornucopia", cover image, *Focus,* March, 2000.

[5] A. K. Dewdney, Four-Dimensional Dementia, *Scientific American,* April, 1986.

[6] M. Kaku, *Hyperspace,* March, 1995.

[7] M. L'Engle, *A Wrinkle in Time,* Dell Publishing Co., New York, NY, 1962.

[8] J. Weeks, *Shape of Space,* Monographs in Pure and Applied Mathematics, Marcell Dekker, Inc., New York, 1985; see [URL: :http://www.geom.umn.edu/locate/SOS].

[9] Pixar Animation Studios, *Geri's Game,* animated short film, 1997.

[10] Disney Animation Studios, *Toy Story,* animated feature film, 1999.

[11] The MAM2000 web site, [URL: :http://mam2000.mathforum.com/].

[12] The Math Forum, [URL: :http://www.mathformum.com/].

[13] Equations for Cornucopia, [URL: :http://mam2000.mathforum.com/612/dimension/cone/equations.html].

Visual Topology and Variational Problems on Two-Dimensional Surfaces

ANATOLY T. FOMENKO, ALEXANDR O. IVANOV, and ALEXEY A. TUZHILIN

The aim of the present paper is to illustrate the modern state of variational calculus in large with several physical and geometrical examples. In spite of the great progress in variational calculus during the last century, some classical variational problems are still far from its final solution. But, as it often happens in Science, the attempts to solve classical problems give rise to beautiful theories that give us an understanding of the problems and, sometimes, lead to unexpected discoveries. As typical examples we have chosen *Minimal Surfaces Theory* and *Extreme Networks Theory* that attract the attention of many scientists during several ages.

Interfaces Between Two Media

The theory of minimal surfaces and surfaces of constant mean curvature is a branch of mathematics that has been intensively developed, particularly recently. On the basis of this theory we can investigate soap films and soap bubbles, interfaces between two media, which occur widely in chemistry and biology, for example, membranes in living cells, capillary phenomena, and so on. Minimal surfaces also turn out to be useful in architecture. Before giving exact definitions, let us consider some examples.

Soap Films and Soap Bubbles

If we dip a wire contour into soapy water, and then carefully lift it out, a soap is left on the contour. For many "young researchers" this is the time to obtain an iridescent bubble by blowing on the film. However, the soap films themselves conceal unexpected properties, which we can immediately see by making a simple experiment. Bend a wire contour, as shown in Fig. 1. This is a so-called Douglas contour. Let u and v denote the distances between the extreme circles of the contour. It turns out that by lifting the contour out of the soapy water differently we can obtain different soap films. Fig. 2 shows some of them (for small $u = u_0$ and $v = v_0$).

Fig. 1

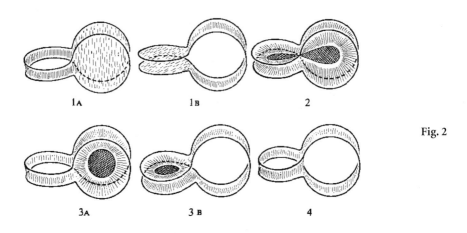

1A   1B   2

3A   3B   4

Fig. 2

If the contour is not deformed in the process of obtaining the film, then as a rule films of type 1a and 1b are formed. A film of type 2 is obtained if at the time of lifting the contour out of the soapy water the left and right circles are kept joined ($u = 0, v = 0$), but after lifting the contour out they are let free. The elastic contour returns to its original position ($u = u_0, v = v_0$), and a film of type 2 remains on it. Films of type 3a and 3b can be obtained similarly, by combining only right (or only left) circles. To obtain a film of type 4 it is sufficient to puncture a film of type 3a by a disc D^2.

We observe that films of type 2 and 3 are not smooth surfaces. They have singularities: a singular point A where four singular edges meet (type 2), or a singular edge S^1 (type 3). Moreover, smooth films of types 1 and 4 have different topological type: a film of type 1 is a two-dimensional disk D^2, while a film of type 4 is a torus with a point deleted (see Fig. 3).

Thus, on a given contour we can, generally speaking, stretch many soap films, and not every film need be a smooth surface. Moreover, on a contour that is a bent circle we can stretch a minimal torus with a point deleted (a film of type 4). How

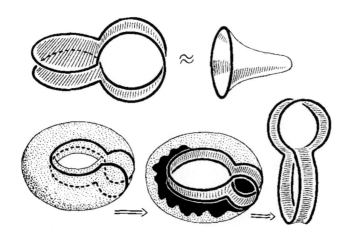

Fig. 3

many soap films can span a given contour? What topological types of films can occur? What singularities can be found on a soap film? These and other questions are considered in the theory of minimal surfaces (see below).

Soap bubbles and soapsuds, that is, a system of soap bubbles, are just as interesting to the researcher. Physical bubbles differ from soap films in that bubbles bound regions of space in which air is under greater pressure than the air outside. If we look closely at the soapsuds, we observe that the singularities at the junction of different bubbles are similar to the singularities of soap films. By carrying out many experiments with the interfaces between two media the Belgian physicist Joseph Plateau (1801–1883) formulated four principles, which describe the possible stable singularities on these surfaces. It turns out that the two types of singularities revealed above – a smooth singular edge at which three sheets meet, and a singular point at which four smooth singular edges meet, each pair of which is spanned by a smooth sheet – are the only possible singularities of stable soap films and soapsuds (Plateau's 1st, 2nd, and 3rd principles). Moreover, in the first case the sheets meet at a singular edge at an angle of 120°, while in the second case the singular edges meet at a point at an angle of about 109°28'16" (more precisely, the cosine of this angle is –1/3), like the four line segments drawn from the center of a regular tetrahedron to its vertices (Plateau's 4th principle). We give some elementary justification of these principles below, first describing the variational principle that underlies soap films and soap bubbles.

The Poisson–Laplace Theorem

Soap films and soap bubbles can be regarded as the interfaces between two homogeneous media in equilibrium. A soap film with boundary locally, that is, in a neighborhood of each of its points, separates two media, air–air, in each of which the pressure is the same. Therefore the total pressure on each small area of a soap film is zero. In a soap bubble the pressure inside is greater than the pressure outside, so the vector of the resultant pressure is directed outwards. This force must be compensated by the forces of surface tension. Since the pressure is always directed along the normal to the interface and is the same in absolute value at all points of this interface because of the homogeneity of the media, the interface is "curved in the mean" identically at all its points. To give a precise meaning to this statement, we need to define the geometric concept of "mean curvature" (for the details see [3], [19]).

Let M be a smooth two-dimensional surface in $\mathbb{R}^3$, suppose that the point P lies on the surface M, and that $N(P)$ is one of the two unit normals to M at P (the vector $N(P)$ is orthogonal to the tangent plane to M passing through the point P; we denote this tangent plane by $T_P M$).

Through P we pass a plane Π containing the $N(P)$. The plane Π intersects M along a curve γ called a *normal section*. The unit vector v tangent to γ at P is called a *direction* of this normal section. Clearly, the vector $-v$ determines the opposite direction of the normal section γ, and the vectors v and $-v$ lie in the tangent plane $T_P M$ (Fig. 4).

Let $k(v)$ denote the curvature vector of γ in the direction v, that is, the acceleration vector at P under motion along γ with unit speed. We note that $k(v) = k(-v)$.

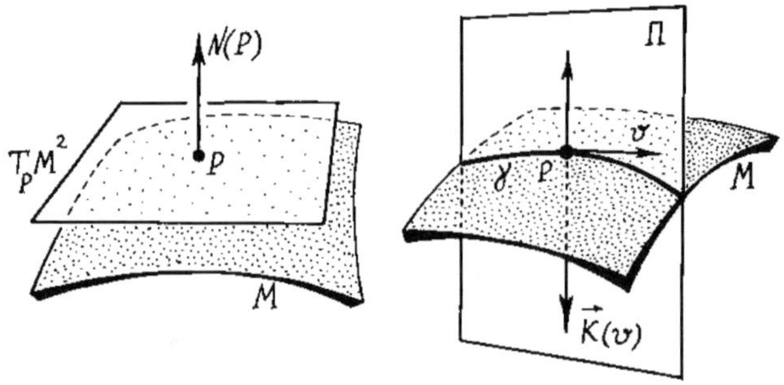

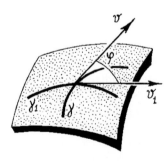

Fig. 4

It is easy to show that the curvature vector $k(-v)$ is collinear with the normal $N(P)$. We define the curvature $k(v)$ of a normal section γ in the direction v with respect to the normal $N(P)$ as the quantity $k(v) = \langle\, k(v), N(P)\rangle$, where the brackets $\langle,\rangle$ denote the standard scalar product of vectors in $\mathbb{R}^3$.

Clearly, the continuous function $k(v)$ takes maximum and minimum values (since $k(v)$ is a function on the (compact) circle S^1 formed by all directions v). These values are called the *principal curvatures* k_1 and k_2 of the surface M at the point P, and the normal sections in which the values k_1, and k_2 are attained are called the *principal sections*.

Definition. The mean curvature H of a surface M at a point $P \in M$ with respect to the normal $N(P)$ is half the sum of the principal curvatures: $H = (k_1 + k_2)/2$.

Let φ denote the angle between the direction v of an arbitrary normal section γ at a point $P \in M$ and the direction of the principal section γ_1 (Fig. 5). If k_1 is the curvature of the principal section Γ_1, and k_2 is the other principal curvature, then by Euler's well-known formula

$$k(v) = k(\varphi) = k_1 \cos^2 \varphi + k_2 \sin^2 (\varphi).$$

Fig. 5

Fig. 6

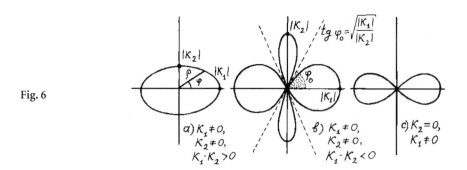

a) $K_1 \neq 0$,
   $K_2 \neq 0$,
   $K_1 \cdot K_2 > 0$

б) $K_1 \neq 0$,
   $K_2 \neq 0$,
   $K_1 \cdot K_2 < 0$

c) $K_2 = 0$,
   $K_1 \neq 0$

To represent more clearly the distribution of the curvatures of the normal sections γ as the angle φ changes, we construct in the plane with polar coordinates (ρ, φ) the graph $\rho = |k(\varphi)|$. We can distinguish the following cases:

a) k_1 and k_2 are nonzero and have the same sign. In this case the graph is an ellipse-like curve with "semi axes" $|k_1|$ and $|k_2|$ (where $|k(\varphi)|$ is maximal and minimal). If $k_1 = k_2 \neq 0$, this "ellipse" degenerates to a circle (Fig. 6a);
b) k_1 and k_2 are nonzero and of different signs. In this case the graph is a "four-leafed rose" (Fig. 6b);
c) one of the curvatures k_i is zero – the "four-leafed rose" degenerates into a "two-leafed rose" (Fig. 6c);
d) both the principal curvatures are zero. The graph is a point (the origin).

It is now clear that if k_1 is not equal to k_2, then there are exactly two principal sections, which are in fact orthogonal to each other. If the principal curvatures are equal, then the curvature of all the normal sections is the same and equal to the mean curvature H.

Remark. It is easy to see that half the sum of the curvatures of any two mutually orthogonal normal sections is constant and equal to H.

Theorem 1.1 (Poisson–Laplace) *Assume that a smooth two-dimensional surface M in $\mathbb{R}^3$ is the interface between two homogeneous media in equilibrium. Let P_1 and P_2 be the pressures in the media. Then the mean curvature H of the surface M is constant and equal to $H = h(P_1\text{-}P_2)$, where the constant $\lambda = 1/h$ is called the coefficient of surface tension, and $P_1\text{-}P_2$ is the difference between the pressures in the media (the resultant pressure).*

Thus, the expression "curved in the mean identically" implies that the mean curvature of the surface is constant. Taking account of what we said above, we can conclude that the mean curvature H of a soap film is zero, $H \equiv 0$, and the mean curvature H of a soap bubble is a constant $\neq 0$. In mathematics surfaces with $H =$ const are called *surfaces of constant mean curvature*. For the case $H = 0$ these sur-

103

faces have a special name – *minimal surfaces* (due to the fact that they locally minimize the area functional). Sometimes they are called *soap films*, and surfaces with $H = \text{const} \neq 0$ are called *soap bubbles.*

Surfaces of constant mean curvature are widespread in nature and play an important role in various research. Thus, for example, surface interactions on the interface between two media determine the character and rate of chemical reactions. Various membranes, such as the ear-drum and membranes that separate living cells, are minimal surfaces. One more example consists of microscopic marine animals – Radiolaria (see [2], [4], [5], [6]).

The Principle of Economy in Nature

In this section we talk about an alternative approach to the description of minimal surfaces and surfaces of constant mean curvature, based on the variational principle (for a more detailed historical survey see [11]).

Optimality and Nature

In 1744 the French scientist Pierre-Louis-Moreau de Maupertuis put forward his famous principle, which has become known as the principle of least action. In 1746 Maupertuis published a paper *The laws of motion and rest deduced from a metaphysical principle.* This metaphysical principle is based on the assumption that Nature always acts with the greatest economy. Starting from this position, Maupertuis draws the following conclusion: if certain changes occur in Nature, then the total action needed to carry out these changes must be as small as possible.

In parallel and independently Leonard Euler in 1744 obtained a strict proof that the principle of least action can be used to describe the motion of a material point in a field of conservative forces such as the motion of the planets around the Sun. Euler also put forward the conjecture that for any phenomenon in the Universe we can find a maximum or minimum rule to which it is subject. This remark appeared in the Appendix to his famous work of 1743 *Methods of finding curves that are subject to a maximum or minimum property,* the first textbook on the calculus of variations.

When in 1746 Maupertuis published his work on the principle of least action he was well aware of Euler's achievement, since he briefly described it in the Preface. Then, however, he added: "This remark ... is a beautiful application of my principle to the motion of the planets", thus asserting his priority.

Euler reacted to this remark by giving up his right of priority, for which he was strongly criticized by certain historians of science. We shall not go into the details of the subsequent keen discussion that developed over priority in the discovery of the principle of least action. We shall only say that other people (König and apparently Leibniz) laid claim to authority and that the discussion was linked to the anthropomorphic understanding of the terms "living force" and "action" and with theology (see [11], [17]).

However, we observe that Maupertuis, who formulated his principle starting from the idea of the perfection of God, tested it on a few examples, but he did not

Fig. 7

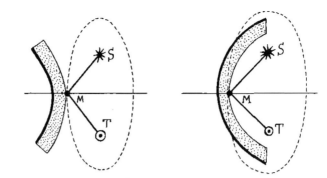

investigate some of them thoroughly. It turns out that the principle of least action is not always true.

Let us consider one of the examples given by Maupertuis – the reflection of light. Here the law of least action leads to the conclusion that a ray of light "selects" from all possible routes from the source to the receiver the one that can be covered in the least time (this rule had already been formulated by Fermat). If light is propagated in a homogeneous medium, then this minimum principle leads to the simpler rule: a ray of light moves along the shortest path joining the source and the receiver.

Consider a spherical mirror situated in a homogeneous medium. Suppose that the source S and the receiver T are symmetrical about some line ℓ passing through the center of the sphere. What characterizes the trajectory SMT of the motion of light emitted from the source S, reflected in the mirror at a point M, and received by the receiver T? In this situation we apply the famous law presented in the work *Catoptrica* attributed to Euclid – the so-called law of reflection. The following brief formulation of it is well known: the angle of the incidence is equal to the angle of reflection.

Figure 7 shows two situations: a convex mirror (Fig. 7a) and a concave mirror (Fig. 7b). In both cases the point M is the point of intersection of the mirror and the line ℓ (we consider the trajectories of the motion of light in the plane passing through the source, the receiver, and the line ℓ). Through M we draw the ellipse with foci at S and T. If M_1 is an arbitrary point lying outside the ellipse, M_2 is an arbitrary point inside it, and M is an arbitrary point on it, then $|M_1S| + |M_1T| > |MS| + |MT| > |M_2S| + |M_2T|$.

Hence it follows that in the case of a convex mirror the trajectory SMT actually has the shortest length, but for a concave mirror this is not always so. In Fig. 7b the source S and the receiver T are symmetrical about the center of the sphere. It is easy to see that any path SM_2T, where $M_2 \neq M$, is shorter than the path SMT.

Thus, whether Nature is most economical or most wasteful depends on the form of the mirror. Citing similar arguments, d'Arcy showed in 1749 and 1752 that the principle of Maupertuis was not clearly formulated, and leads to incorrect assertions.

Nevertheless, the idea of optimality of the phenomena of Nature plays an important role in physics. The mathematical formulation of this idea has given birth to the calculus of variations, the founders of which are usually taken to be Lagrange and three Swiss mathematicians from Basel: the brothers Johann and Jakob Bernoulli and a student of Johann Bernoulli – Leonard Euler.

Minimal Surfaces and Optimality

It turns out that the forms of soap films are also optimal in a certain sense, namely the corresponding minimal surfaces are the extremals of the area functional. Let us explain this statement. For this we consider a soap film covering a given contour. Surface tension leads to the film tending to take up a form with the least possible surface energy (of course, this is an approximation, which nevertheless works well in practice). Since the surface energy is directly proportional to the surface area, as a result of minimizing the surface energy the area of the soap film is least compared to the areas of all sufficiently neighboring surfaces covering the given contour.

Thus, soap films are local minima of the area functional. However, minimal surfaces, that is, surfaces of zero mean curvature, need not minimize the area even among all neighboring surfaces (with given boundary). To explain this we note that the concept of "neighboring surfaces" can be defined in different ways. We shall understand by a neighboring surface of a given surface M one obtained by a small deformation of M, leaving the boundary ∂M fixed. For a zero-dimensional surface M, that is, when M is a point, a small deformation is a small displacement of M. When the dimension of M is at least one, we can define two types of small deformations: a deformation with small amplitude, that is, when each point of M is displaced not far from its original position, and a deformation with small support, when only a sufficiently small region of M undergoes a deformation; the closure of such a deformed region is called the support of the deformation. A deformation with sufficiently small support always increases the area of the minimal surface, so for such deformations the minimal surfaces do minimize the area. For deformations with small amplitude this is not so. Nevertheless, for such deformations minimal surfaces are extremals (critical points) of the area functional.

It is easy to show that the (finite) area of any surface in $\mathbb{R}^3$ can always be increased by an arbitrarily small deformation of this surface, so no surface can be a local maximum for the area functional. It is well known that critical points other than a local minimum and a local maximum are called saddle points. Minimal surfaces corresponding to saddle points of the area functional (for deformations with small amplitude) are said to be unstable. If we have succeeded in creating a soap film that has the form of an unstable minimal surface, then fluctuations of this film that are small in amplitude, which always exist in the real world, would instantaneously lead to its collapse – the film constructed would turn out to be unstable.

Thus, minimal surfaces are critical points of the area functional. It turns out that the converse is true: a surface M that is a critical point of the area functional (considered on the space of all possible surfaces close (in amplitude) to M and having the same boundary ∂M) is minimal, that is, it has zero mean curvature.

Surfaces of constant mean curvature are also extremals of a certain functional. They can also be obtained as extremals of the area functional if we restrict the possible deformations. As an example we consider a soap bubble. If we blow on it, the film, sagging in one place, will swell at another place in such a way that the volume of the region inside the bubble is unchanged. This observation is the basis of the definition of a surface of constant mean curvature from the viewpoint of the variational principle. For a closed surface bounding a region in $\mathbb{R}^3$, as admissible deformations we consider only those that preserve the volume of the region bounded by this surface. The condition that the volume is preserved can be described in yet another way. For this we define the function of change of volume of a region V bounded by a surface M under a deformation M_t of this surface. For each $t = t_0$ we consider the totality of regions included between the surfaces M and M_{t_0}. From the total volume of those that lie outside V we subtract the total volume of those that lie inside V, and we call the resulting number the change in volume at the instant $t = t_0$. Varying t_0, we obtain a function which is called the *change of volume function*. Clearly, the regions lying inside and outside V are on opposite sides of M, and so they can be defined without the use of V.

This observation enables us to define the change of volume function for a deformation M_t of an unclosed surface M (such as the soap film bounded by a wire contour or a hemisphere having the equator as its boundary), but in this case we must require that the deformation M_t is fixed ($= 0$) on the boundary ∂M of the surface M.

We say that a deformation M_t of a surface M (fixed on ∂M if $\partial M \neq 0$) preserves the volume if the change of volume function constructed from this deformation is identically zero. It turns out that *surfaces of constant mean curvature are critical points of the area functional restricted to the space of deformations that preserve the volume.*

Since surfaces of nonzero constant mean curvature are not minimal surfaces, they are also not critical points of the area functional considered on the space of all possible deformations (fixed on the boundary), not only those that preserve the volume. Thus, restriction of the space of admissible deformations naturally leads to an increase in the number of surfaces that are critical points of the area functional considered on this space.

A description of minimal surfaces as extremals of the area functional proves very useful. For example, this approach has given the possibility of solving the so-called Plateau problem, which consists in the following: for any contour, among all surfaces of given topological type that bound it, is there a surface of least area? A positive answer to this problem in the case when the contour is a simple rectifiable Jordan curve (it has finite length) and the surface has the topological type of the disk D^2 was obtained in 1928 by the young American mathematician Jesse Douglas. However, his proof turned out to be incomplete, and up to 1931 his paper had not appeared in print. At about the same time a solution of Plateau's problem, obtained by the Hungarian mathematician Tibor Rado, was published. In the following decades Jesse Douglas also solved a number of other problems that arose in the theory of minimal surfaces. In particular, the powerful mathematical technique that he developed enabled him to prove the existence of minimal surfaces of high genus spanning one or finitely many contours. For his achievement Dougals was awarded in 1936 the highest prize in mathematics – the Fields Medal.

Let us mention that in high dimensions Plateau problem is even more complicated. To find a solution it is necessary to generalize the concepts of a surface and a boundary. We can not go into details here and only list the names of scientists who had solved the generalized Plateau problem in 60–70th: F. J. Almgren, E. De Giorgi, H. Federer and W. H. Fleming, A. T. Fomenko, Dao Trong Thi (see details for example in [2]).

In conclusion of Section 2 "The principle of Economy in Nature" we give the one-dimensional version of Plateau's problem, the so-called Steiner problem, and show how from its solution there follows a proof of Plateau's empirical principles, which describe all possible singularities of soap films of stable minimal surfaces. In Section 3 "Extreme Networks" we discuss the one-dimensional case in more details.

Steiner's Problem

Let us begin with the simplest case. Suppose we need to connect the towns A, B, and C by a system of roads, that there are no obstructions, and that we are free to construct the roads where we like. Suppose that the region in which the towns lie is flat. The problem is to find a system of roads of least length. In mathematical language this means the following: for three given points A, B, and C lying in a plane, to find a point P and paths joining P to A, B, and C so that the total length of these paths shall be as small as possible. Since a line segment is the shortest path between its ends, the required paths are line segments PA, PB, and PC. It remains to choose P in an optimal way.

It turns out that the solution depends on the relative positions of A, B, and C. If all the interior angles of the triangle ABC are less than 120°, then the required point P is uniquely determined from the condition that the angles $\widehat{APB}$, $\widehat{BPC}$, and $\widehat{CPA}$ are equal (they are thus equal to 120°). But if one of the angles of the triangle ABC, say the angle at C, is at least 120°, then P coincides with C (see Fig. 8).

The proof of this assertion is based on some elementary geometrical lemmas.

Lemma 2.1 (Heron's theorem) *Suppose that points A and B do not lie on a line a. Then among all the points P of the line a the point $P = P_0$ such that $|AP| + |BP|$ is as small as possible is uniquely determined from the following condition: the angle between AP_0 and the line a is equal to the angle between BP_0 and the line a (Fig. 9).*

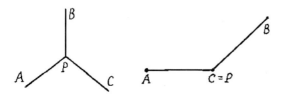

Fig. 8

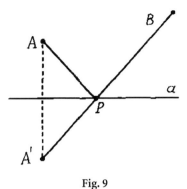

Fig. 9

If A is a source of light, B is a receiver, and a is a mirror, then Heron's law can be regarded as a special case of the law of reflection (see above).

Lemma 2.2 *Suppose that the points A and B are the foci of an ellipse, and that the line a touches this ellipse at a point P. Then the angles between the line a and the segments AP and BP are equal (Fig. 10).*

It follows that if we put a source of light at one focus of an elliptical mirror, then all the rays collect at the other focus. Moreover, in elliptical billiards the ball always goes either outside the foci, or through the foci, or between the foci (see Fig. 10).

109

To prove Lemma 2.2 it is sufficient to observe that the sum of the distances from any point outside an ellipse to its foci is greater than the sum $|PA| + |PB|$ (since P lies on the ellipse).

Now suppose that P is a solution of Steiner's problem for a triangle ABC in which all the interior angles are less than $120°$. Through P we draw the ellipse whose foci are A and B. Through P we draw the tangent a to this ellipse (see Fig. 11). It is easy to see that CP is perpendicular to a. Therefore from Lemma 2.2 we see that $\widehat{APB} = \widehat{BPC}$. Similarly, $\widehat{BPC} = \widehat{APB}$. Thus, all the angles at P are equal to each other, and therefore equal to $120°$. Now it is easy to construct the required point P and to see that the solution is unique.

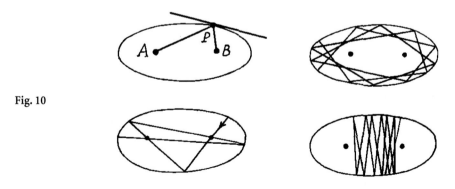

Fig. 10

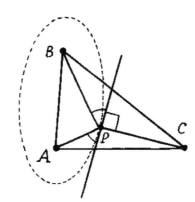

Fig. 11

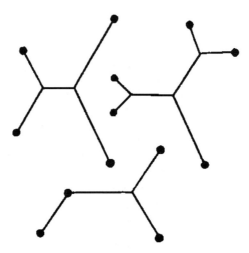

Fig. 12

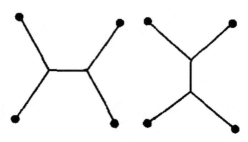

Fig. 13

If instead of three points A, B, and C we take any finite number of points, we obtain the generalized Steiner problem: it is required to join all these points by a finite system of curves of least length. This problem can be restated as follows: how do we join n towns by a network of roads with the least expense? From the combinatorial point of view the solution of this problem is a combination of two networks, described above for the case $n = 3$. Here are some examples (Fig. 12).

We observe that the solution of the generalized Steiner problem is not unique. For example, if four points are vertices of a square, we can obtain two symmetrical solutions (Fig. 13). We note that a system of paths of least length joining n points of a plane is called a shortest network (in the plane), or Steiner minimal tree.

The generalized Steiner problem in the plane has still not been completely solved, so an experimental "solution" is of interest. Take two glass or transparent plastic sheets, place them in parallel planes, and join them by pieces of wire of the same length, equal to the distance between the sheets. Clearly, all the pieces of wire are parallel to one another and perpendicular to the sheets.

If we dip this configuration into soapy water, and carefully lift it out, then between the sheets there will be a soap film whose boundary consists of two parts: the set of joining wires and the set of "traces" which the film leaves on the sheets. We note that in accordance with the minimum principle the film is at an angle of 90° to each sheet. Moreover, this film consists of perpendicular sheets of planar rectangles that adjoin each other along the singular edges (see Fig. 14). If a singular edge is not a joining wire, then in accordance with Plateau's principles exactly three rectangles meet on it at angles of 120°. We observe that the area of the resulting soap film is equal to the total length of the path joining the points where the wires are fastened (the "trace" of the soap film on a sheet) multiplied by the distance between the sheets. If the film has least area among all films with a partially free boundary consisting of the joining wires (rigid boundary) and the "traces" on the sheets (the hypothesis of the existence of such a film is called Plateau's problem with partially free boundary), then the corresponding "trace" on a sheet is a solution of the generalized Steiner problem for the configuration given by the fastening points.

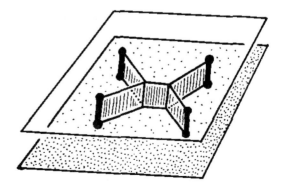

Fig. 14

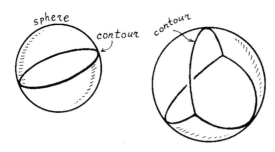

Fig. 15

Let us now turn to Plateau's principles, which describe all possible singularities of stable minimal surfaces. On a soap film we choose an arbitrary point P. We take a smaller and smaller neighborhood of P and blow it up to the same size. In the limit all the surfaces that join at P become planar, and the singular edges become segments of straight lines. Clearly, after such an enlargement the resulting fragment of stable film will also be stable. Now consider a sphere S^2 with center at P. Its intersection with the planar configuration we have constructed is a networks on the sphere S^2. Clearly the curves of intersection are parts of great circles of S^2. Moreover, if l denotes the length of the network, and r is the radius of the sphere S^2, then the area s of the part of the planar configuration inside the sphere is equal to $s = lr/2$. Therefore from the stability of the film it follows that at each node of the network only three arcs can meet at angles of 120° (otherwise by a small deformation we could lower the length of the network, and hence the area of the film).

The following natural problem arises (the so-called Steiner problem on the sphere): to describe all possible networks on the sphere consisting of arcs of great circles meeting at each vertex of the network three at a time at equal angles (of 120°). In contrast to the planar Steiner problem, the spherical problem has been completely solved. It turns out that there are exactly ten such networks, drawn in Fig. 15–22. A more careful analysis shows that only three of these ten networks (the first three in Fig. 15 and 16) correspond to configurations that minimize the area. Fig. 23, 24, and 25 show soap films stretched on contours corresponding to minimal networks on a sphere. Only the first three of them are cones corresponding to the local arrangement of soap films described in Plateau's principles. This observation is a physical "proof" that only in the first three cases are the cones absolutely minimal, that is, they correspond to singularities occurring in stable soap films. Let us mention that a mathematical proof of Plateau's principles was given by J. Taylor [18].

Extreme Networks

The present Section is devoted to a new branch of mathematics studying branching solutions (*extremals*) of one-dimensional variational problems. This branch called *Extreme Networks Theory* appeared during attempts to understand how one can solve the generalized Steiner problem (see above).

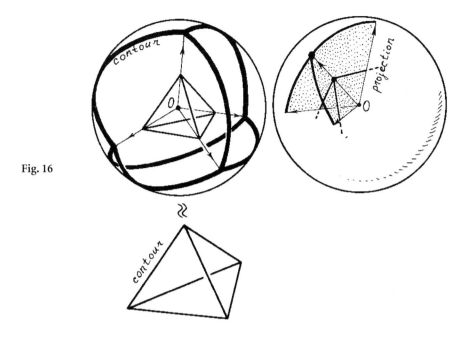

Fig. 16

We recall that Classical One-Dimensional Variational Problem can be stated as follows: to describe extremals of a variational functional defined on a space of curves joining a pair of fixed points in an ambient space. The theory of such extremals (so called *extreme curves*) is well elaborated and presented in standard university courses such as Optimal Control, Differential Geometry, and Classical Mechanics. Let us assume now that we are given not with two but with three or more points of ambient space. The following natural questions arise: How is it possible to generalize the theory to this case? How is it possible to generalize the notion of an extreme curve? In such a way we come to an idea to introduce a new object, *networks,* which can be represented as unions of a finite number of curves into consideration.

Extreme networks for elementary variational functionals such as the length functional appeared in several ways in different theoretical and applied investigations. However, the problem was not systematically studied, in spite of its classical statement. Apparently, first problems of that type appeared in works of French mathematicians, Gergonne, Clapeyron and Lamé are among them. C. Gauss was also interested in such problems. In his letter to Schuhmacher, he brought up a question how to build the shortest system of roads joining four German towns: Hamburg, Bremen, Hannower, and Braunschweig. General problem of finding of the shortest networks joining a given finite set of points in the plane was stated by Jarnik and Kössler, see [16]. Later on, this problem became widely well-known as *Steiner problem* due to the famous book of Courant and Robbins (1941), see historical reviews in [1], [13], [14], or [15]. Notice that Steiner problem has several sorts, among them there is so-called *rectangular* Steiner problem consisting in description of the planar shortest networks whose edges are polygonal lines with parallel to

113

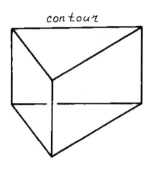 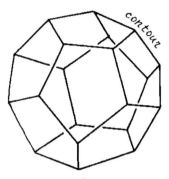

con tour

contour

Fig. 17

(minimal film ≠ cone)

coordinate axes links. The later problem appears naturally under arrangement of wires on printed circuit cards, see [11], and [13].

Thus, the idea of networks studying appeared several ages ago, and the interest to Steiner problem did not fade away as times goes by. Moreover, today a systematic studying of networks fulfilling some natural conditions, especially some extremality conditions, becomes even more important due to rapid development of networks of different kinds and levels. As an example we can speak about transportation networks, computer networks, networks prescribed by the structure of complex chemical molecules, such as DNA, etc.

What is the difference between the classical approach and the approach of extreme networks theory?

Firstly, main classical works concerning extreme networks are devoted to the studying of the shortest networks, the considered functional is usually either Euclidean length functional in a vector space (in the plane for the most part), or Manhattan length[1] functional (also in the plane mainly). But extreme networks theory supposed to give a common approach to investigate networks extreme with respect to an arbitrary functional in an arbitrary ambient space, say in a smooth manifold, or in a manifold with singularities such as a surface of polyhedron or an A. D. Alexandrov space, etc.

Secondly, the local structure of the shortest networks, that is, the structure of sufficiently small neighborhoods of the network points, is determined rigid enough. But if we consider extreme networks instead of the shortest ones then several possibilities appear. Notice that extremality is the networks property which characterizes the behavior of the functional under small deformations of the network. In the case of curves, the deformations are defined in more or less unique and natural way. In the case of networks one can consider several essentially different types of deformations. But the following two types are most important: deformations preserving the structure of the network, and deformations which can split some vertices of the network (see an example of such splitting of the vertex in Fig. 26). Notice that the network Γ, spanning the vertices of the

---

[1] Recall that Manhattan lengths of vector equals to the sum of the lengths of its projections to coordinate axes.

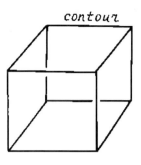

contour

Fig. 18

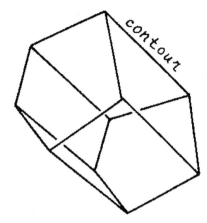

contour

(*minimal film ≠ cone*)

Fig. 19

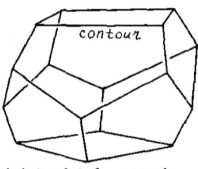

contour

(*minimal film ≠ cone*)

Fig. 20

115

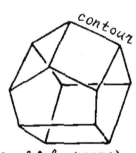

contour

(*minimal film ≠ cone*)

Fig. 21

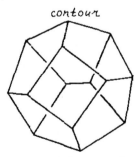

contour

(*minimal film ≠ cone*)

Fig. 22

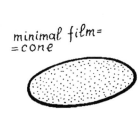

minimal film=
=cone

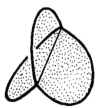

minimal
film = cone

Fig. 23

Fig. 24

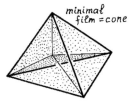

minimal
film =cone

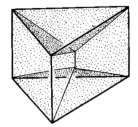

minimal film ≠ cone

116

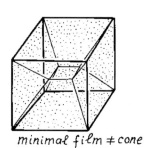

Fig. 25

minimal film ≠ cone

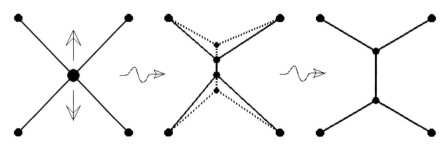

Fig. 26. Splitting of the vertex

square and having exactly one interior vertex v located in the center of the square, is an extremal for the length functional with respect to the deformations preserving the network's structure. However, the deformation splitting the vertex v as it is shown in Fig. 26 decreases the length of the network Γ. The velocity of this decrease is not zero, thus the network Γ is no longer extreme with respect to such deformations.

The networks which do not change their structure under deformations are called *parametric networks*. The networks which allow the vertices splitting, that is, the networks whose structure can change under deformations are called *networks-traces*. (Formal definitions of these types of networks can be found in [15]).

Thirdly, by their own nature the networks possess both geometrical and combinatorial characteristics, they have both continuous and discrete properties. Therefore, creation of extreme networks theory initialized the development of new methods which do not have analogue in classical theory. The technique appeared combines methods of differential geometry and variational calculus on the one hand, and methods of discrete geometry and combinatorial analysis on the other hand. Besides, the authors hope that the continuous-discrete nature of networks could give a new way of thinking about problems dealing with geometrical quantization (in the most wide meaning of the words). Namely, properties of extreme networks (with respect to an appropriate variational functional) contain information about both continuous (geometrical) and discrete (quantum) characteristics of ambient space.

Below we present several spectacular geometrical results obtained in Extreme Networks Theory. An interested reader can find details in [14] or [15].

Possible Structure of Planar Extreme Network Depends on Geometry of its Boundary

We consider planar binary trees (that is, planar connected graphs without cycles, whose vertices degrees equal either 1 or 3) that are extreme with respect to the standard Euclidean length functional. Such networks are usually called *local minimal 2-trees* or *full Steiner minimal trees*, and it is well-known, see for example [13], that each shortest tree in the plane can be naturally represented as a union of such networks. Moreover, the most present-day algorithms constructing shortest on an enumeration of all such possible local minimal 2-trees. Therefore, each *a priori* elimination of possibilities can be applied to improve the existing algorithms solving Steiner problem. We give such an effective restriction in terms of geometry of the boundary set.

Recall the well-known partition of an arbitrary finite non-empty subset M of the plane into convexity levels. We assign all the points of M that lie on the boundary of the convex hull of M to the *first convexity level* M^1 of M. Notice that M^1 is not empty. We remove M^1 from M. If the resulting set is not empty we transform it by the same procedure. Namely, all the points of M that lie on the boundary of the convex hull of $M \setminus M^1$ are assigned to the *second convexity level* M^2 of M. We continue this process until all the points of M lie on some convexity level. The resulting partition $M = \sqcup_{i=1}^{k} M^i$ is called the *partition into convexity levels* and the set M^i is called the *ith convexity level of M*. If $M = \sqcup_{i=1}^{k} M^i$, then we say that M has *k convexity levels*. We denote by $k(M)$ the number of convexity levels of the set M. Note that M lies on the boundary of its own convex hull if and only if it has a single convexity level, which coincides with M itself.

Now we define a characteristic of a planar binary tree, that introduced in [14]. Let Γ be a planar 2-tree, and let a and b be some edges of Γ. Choose the path $\gamma \subset \Gamma$ connecting a and b, and let it be oriented from a to b.

Let the plane be oriented. Then the notions "left" and "right" are defined. Let us move along γ from a to b. During this motion, in each interior vertex of γ we turn either to the left or to the right (we omit rigorous definitions because they are obvious but cumbersome). We denote by tw(a, b) the difference between the numbers of left and right turns. The tw(a, b) is called *the twisting number* of an ordered distinct pair (a, b). We put tw$(a, a) = 0$.

Definition. The *twisting number* tw(Γ) of a planar 2-tree Γ is the maximum among the twisting numbers of all the ordered pairs of edges of Γ: tw$(\Gamma) = $ max$_{(a,b)}$ tw (a,b).

The following result demonstrates the connection between the twisting number of a local minimal 2-tree and the number of convexity levels of its boundary.

Theorem 3.1 *Let (Γ) be a planar local minimal 2-tree spanning a finite subset M of the plane. Then* tw (Γ) $k(M) \leq 12$ $(k(M) - 1) + 5$.

If $k(M) = 1$, then the following inverse statement holds

Theorem 3.2 *Let Γ be a planar binary three with* tw $\Gamma \leq 5$. *Then there is a local minimal tree Γ_m spanning a finite subset M of the plane, such that Γ_m is planar equivalent to Γ and $k(M) = 1$, that is, M lies on the boundary of its own convex hull.*

Possible generalizations of these results for the case of so-called weighted length functional and the corresponding results in planar graph theory are discussed in [15].

Infinitely Many Different Closed Extreme Networks on a Flat Torus

In Section 2 "The Principle of Economy in Nature" we discussed closed minimal networks in the standard sphere. We recall that up to isometry there are exactly 10 such networks. But it turns out, that on closed flat surfaces the situation is quite other. A. Ivanov and A. Tuzhilin in collaboration with I. Ptitsina obtained a complete description of closed local minimal networks on flat tori and flat Klein bottles. We omit the precise statements (the details can be found in [14]), and include only several non-trivial geometrical corollaries.

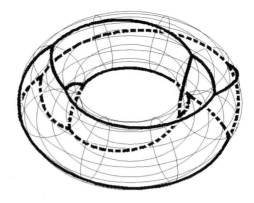

Fig. 27. Minimal network on torus

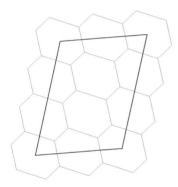

Fig. 28. Topology of minimal
network on torus

119

Theorem 3.3 *In each closed flat surface, that is, in each flat torus and in each flat Klein bottle, there are infinitely many non-equivalent closed local minimal networks. Each closed minimal network on a closed flat surface can be non- trivially deformed in the class of minimal networks.*

Theorem 3.4 *Any sufficiently small deformation of a closed flat surface through the class of homeomorphic flat surfaces does not destroy a closed local minimal network in it.*

Let us remark, that a classification of closed local minimal networks in closed surfaces of constant negative curvature is not obtained still. Some examples of such networks can be found in [14]. But the following general result was proved by M. Pronin who has elaborated an analogue of Morse index theory for minimal networks.

Theorem 3.5 *A closed local minimal network on a surface of strictly negative curvature is rigid, that is, it can not be deformed through the class of local minimal networks.*

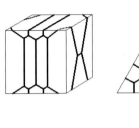

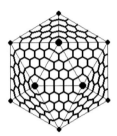
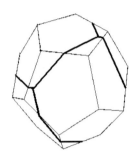

Fig. 29. Closed local minimal networks in surfaces of Platonic bodies

Closed Extreme Networks on Platonic Bodies' Surfaces

Another interesting application of the extreme networks theory appears if we consider local minimal networks in the surfaces of polyhedra. Let us notice, that the surface of each polyhedron can be considered as a flat surface with point-wise singularities corresponding to the vertices of the polyhedron. Networks on polyhedra appears in applications, and first of all in computer science, due to the well-known fact that all surfaces are simulated as some close polyhedra. The detailed discussion of this topic can be found in [14]. Here we mention just one result concerning the most famous polyhedra – so-called Platonic bodies (this result was obtained by A. Ivanov and A. Tuzhilin in collaboration with T. Pavlyukevich (Anikeeva)).

Theorem 3.6 *In the surface of each Platonic body there are infinitely many non-equivalent closed local minimal networks, see Fig. 20.*

Extremality Differs from Local Minimality in Manhattan Plane

It is well-known that networks extremality criteria for Euclidean length functional have local nature, that is, extremality is equivalent to a special structure of small pieces of the network. It turns out that in general case it is not so. As an example we consider networks which are extreme with respect to so-called Manhattan length. We recall that the *Manhattan length of a vector in* $\mathbb{R}^2$ is defined to be equal to the sum of the lengths of its projections onto the coordinate axes. We assume here that a standard basis is fixed in $\mathbb{R}^2$.

Extremals of the Manhattan length functional form an important class of extreme networks. The first works devoted to the investigation of shortest networks with respect to Manhattan length appeared in 60th, see [8], because of rapid

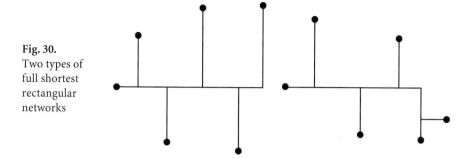

Fig. 30.
Two types of full shortest rectangular networks

development of electronics and robotics. The interest to the Manhattan length emerged in view of that the conductors on printed circuit boards have, as a rule, the form of polygonal lines composed from horizontal and vertical straight segments, therefore their Manhattan length and their Euclidean length are the same. A similar situation takes place in robotics. Apparently, the first systematic investigation of shortest networks with respect to the Manhattan length (so-called *shortest rectangular trees*) was undertaken by Hanan, see [10]. Hanan described several important general geometrical properties of such networks. In particular, Hanan proved that there is a shortest rectangular tree, which is a subset of so-called *Hanan lattice,* that is, the set of all vertical and horizontal straight lines passing through the boundary points. We notice that, generally speaking, the edges of a shortest rectangular tree can be chosen in many ways without changing the length of the tree, of course. However, starting from Hanan, see [10], the tradition approach appeared to chose the edges of such trees as polygonal lines with links parallel to coordinate axes.

Ten years later Hwang, see [12], described possible structure of shortest rectangular trees under assumption that the given boundary set can be spanned by a *non-degenerate* shortest tree Γ_0. The latter means that the degrees of all boundary vertices of the tree Γ_0 equal to 1. In particular, the tree Γ_0 has no vertices of degree 2. Hwang proved that in this case the shortest tree has one of two possible forms shown in Fig. 30.

However, nobody succeeded in finding an effective algorithm constructing shortest rectangular trees. The reason of these failures was explained by M. Garey and D. Johnson [9]. Namely, they proved that the problem of constructing of a shortest rectangular tree is *NP*-complete, that is, most likely there is no polynomial algorithm to solve this problem. Due to this fact the studying of geometry of the shortest rectangular trees becomes even more topical.

As we have already mentioned above, the Manhattan length gives us an example of the functional for which networks extremality criteria is not local even under an assumption that all the edges of the network considered are shortest segments joining the corresponding vertices. Consider the following example: the edges of the network Γ depicted in Fig. 31 are shortest curves, however, this network is not critical. In Fig. 31a deformation which linearly decreases the length is shown.

121

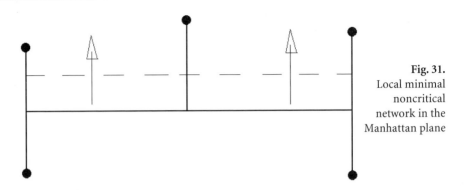

Fig. 31.
Local minimal
noncritical
network in the
Manhattan plane

An interested reader can finnd a criterion of extremality of a tree in Manhattan plane in [15]. This criterion was obtained by A. Ivanov and A. Tuzhilin in collaboration with Höng Van Le. Let us mention that the Manhattan length is a particular case of the length in a normalized space. In [15] we give extremality criteria for networks-traces and for parametric networks in arbitrary normalized space.

References

[1] D. Cieslik, *Steiner Minimal Trees,* Kluwer Academic Publishers, Dordrecht-Boston-London, 1998.

[2] Dao Chong Thi, A. T. Fomenko, *Minimal Surfaces and Plateau Problem,* Nauka, Moscow, 1987 (in Russian). (Engl. translation Amer. Math. soc., Providence, RI, 1991.)

[3] B. A. Dubrovin, A. T. Fomenko, S. P. Novikov, *Modern Geometry,* Nauka, Moscow, 1986. English translation: Part 1, GTM 93, 1984; Part 2, GTM 104, 1985, Springer-Verlag, New York.

[4] M. Emmer, *Bolle di sapone: un viaggio tra matematica, arte e fantasia,* La Nuova Italia, Firenze, 1991.

[5] M. Emmer, *Soap Bubbles in art and science,* M. Emmer, ed., *The Visual Mind,* MIT press, 1993.

[6] M. Emmer, *Soap Bubbles,* video, english version, 27 minutes, Rome (1984).

[7] A. T. Fomenko, and A. A. Tuzhilin, *Elements of geometry and topology of minimal surfaces in three-dimensional space,* Translations of Math. Monographs, AMS, v. 93, 1992.

[8] R. L. Francis, *A note on the optimum location of new machines in existing plant layouts,* J. Indust. Engrg., v. 14, pp. 57–59, 1963.

[9] M. R. Garey, and D. S. Johnson, *The Rectilinear Steiner Problem is NP-Complete,* SIAM J. Appl. Math., v. 32, pp. 826–834, 1977.

[10] M. Hanan, *On Steiner's Problem with Rectilinear Distance,* SIAM J. Appl. Math., v. 14, pp. 255–265, 1966.

[11] S. Hildebrandt, and A. Tromba, *The Parsimonious Universe,* Springer-Verlag, New York, 1996.

[12] F. K. Hwang, *On Steiner minimal trees with rectilinear distance,* SIAM J. of Appl. Math., v. 30, pp. 104–114, 1976.

[13] F. K. Hwang, D. Richards, and P. Winter, *The Steiners Tree Problem,* Elsevier Science Publishers, 1992.

[14] A. O. Ivanov, and A. A. Tuzhilin, *Minimal Networks. The Steiner Problem and Its Generalizations,* CRC Press, N.W., Boca Raton, Florida, 1994.

[15] A. O. Ivanov, and A. A. Tuzhilin, *Branching Solutions to One-Dimensional Variational Problems,* World Scientific Publ., 2001.

[16] V. Jarnik, and M. Kössler, *O minimalnich grafeth obeahujicich n danijch bodu,* Cas. Pest. Mat. a Fys., v. 63, pp. 223–235, 1934.

[17] L. S. Polak, *Variational principals in mechanics,* Gosudarstv. Izdat. Fiz.-Mat. Lit., Moscow, 1959 (in Russian).

[18] J. E. Taylor, *The structure of singularities in soap-bubble-like and soapfilm-like minimal surfaces,* Ann. Math., v. 103, pp. 489–539, 1976.

[19] J. A. Thorpe, *Elementary topics in differential geometry,* Springer-Verlag, New York, 1979.

Metarealistic Rendering of Real-time Interactive Computer Animations[1]

GEORGE K. FRANCIS

Apologia

Two decades ago, when I coined the acronym RTICA (for Real-Time Interactive Computer Animation), it was wishful thinking that we should be able to do such a thing in the classrooms, in our student labs, in our offices, at home! By the end of that decade Silicon Graphics Iris workstations had made all this possible. By 1992, the C in the acronym could be expanded to CAVE, the recursively expanding acronym for a new technology [12] in immersive virtual environments. Along the way, what at first was mostly a manner of speaking has matured into a reusable vocabulary.

My acronym is intended to be pronounced, not spelled out. Try it! Make it sound like *articae:* many artful things. The A also expands variously to Animation, Animator, Animatrix. And, *animation* itself has some welcome double meanings: the process of bringing to life, and that which has been so enlivened. Similarly, the *animator,* like the *editor,* is both the human operator and the graphic tool used to produce the animation. There is, as yet, no such word as *animatrix.* Perhaps there should be: for the tenth muse, who inspires us to write computer programs to make pictures come alive.

125

In this essay I shall explore computer animation from the viewpoint of a mathematician for whom drawing mathematical pictures came as a late vocation, and for whom teaching what he learned became the principal arbiter for many choices and decisions. I hope this attempt to formalize my *axioms* (if only to rationalize my ideosyncrasies) will at least amuse the reader.

First, I draw a sharp distinction between photorealistic rendering of computer images and a species of non-photorealistic rendering. This new discipline [33, 23, 1, 29] treats computer simulations of all graphics media other that photography.

Next, I choose a particular family of mathematical phenomena to illustrate, namely several kinds of *homotopies.* These can be classified according to the difficulty of rendering them as real-time interactive animations. Two related areas of mathematics are particularly hard to imagine and therefore good subjects for

---

[1] *An abridged version of this paper appeared in the proceedings of the MOSAIC 2000 Conference, Seattle, 2000. The real-time interactive computer animations mentioned herein, were presented at the conference* Matematica e Cultura: Arte, Tecnologia, Immagini, *Bologna, 2000 [14]*

metarealistic rendering. They are 3-dimensional non-Euclidean geometry, and the visualization of processes extended in 4 isotropic geometrical dimensions.

Finally, I adopt a more polemical turn of speech which touches on the rightful place of mathematical illustration in the brief history of computer graphics. To maintain focus, I trace the evolution of my metarealistic enterprise, *illiView,* as an example of technology driven research.

Metarealism versus Photorealism

There is no ambiguity in the meaning of photorealism in computer graphics. In 1727 Johann Heinrich Schulze discovered the photographic properties of silver-halide [2, 13ff]. Daguerre, 1839, received a life-long pension for his discovery from the French government. After two and a half centuries of photography and over a century of movies to fall back on, we have no difficulty in deciding to what degree a computer animation really looks and acts *real.* We have an even longer collective pictorial experience, formed in the eons during which artists developed (hardly photorealistic) renderings of every conceivable scene and action. By and large – and exceptions are worth pondering elsewhere – artists drew on the real world. Their pictures used real and familiar objects even when the intended meaning was supernatural, spiritual, or just fanciful. Even impressionism, which Webster [41] defines as "a type of realism the aim of which is to render the immediate sense impression of the artist apart from any element of inference or study of the detail," and subsequent abstract styles (by-and-large!), mean to evoke an emotional or intellectual response which the viewer might have had on looking at the actual model for the picture.

Not so when we use computer graphics to render abstractions that *do not* have concomitant realities. Lately there has been a welcome resurgence of the fascination with the fourth *geometrical* dimension. In the latter half of the nineteenth century there was some popular interest in worlds of four or more isotropic spatial dimensions. Two generations of relativistic science and science fiction has made it difficult to persuade college freshmen and the general public alike that *the* fourth dimension need not invariably be time [4, 7, 25, 28]. Hypercubes rotating in 4-space are found in screen-savers and are routinely assigned as machine problems in computer science courses. Since the advent of computer animation and virtual reality, non-Euclidean geometry has come into fashion [24, 18, 37, 42]. Java applets for drawing lines in the hyperbolic plane and on a spherical surface can be found on the web. Instructive excursions into special and general relativity, molecular biology, cosmology etc. all benefit from particularly *non-realistic* simulations which bend the laws of physics, use stick and balls to model membranes, enlarge planetary diameters a thousandfold, and travel through the cosmos faster than light. These cinematic fictions are necessary so that we can see anything at all on the screen, and to animate it in real-time.

This enterprise deserves its own name, and I propose *metarealistic rendering,* fully aware of the hazard of coining jargon. The literature is littered with multiple terms competing for the same fuzzy meaning, and derelict corpses of splendid terms that nobody uses. But this is not a synonym for what is generally called *virtual reality*. The virtual in VR just means that we use novel projections (stereo, for example) and input controls (head-tracking, for example) to convincingly evoke the illusion of experiencing something real, much as in a dream. The prefix *meta,* having the innocent meaning of "between, with, after, along side of " (Webster) also has less desirable connotations, such as *metaphysical, metaphorical and metamorphosing.* But recall that Aristotle's book on the essential nature of reality merely followed his book on its physical reality. Only later did *metaphysics* suggest something transcendent, philosophical, theological. Similarly *metaphor* has become bent out of shape in computing. No longer just the "poetic reuse of a word," it seems to mean the imposition of a familiar cognitive structure on an unfamiliar one. For example, the *desktop metaphor* trades on our familiarity with desktop furnishings, but then has us use a *mouse* (uncommon on desktops) to move pictures representing files but blasphemously called icons, about the computer screen, and later, to eject a floppy by depositing it in a trash can (a dubious metaphor!). Finally, in computing jargon, the transformation of the physical characteristic of an object, literally its *metamorphosis* (meta = trans, morphe = form), has lost the prefix to become just *morphing.* So, having demythologized some undesirable connotations, let us briefly consider whether the distinction between *photorealistic* and *metarealistic rendering* deserves a place in the philosophy of technology.

The photorealist has this significant advantage over a metarealistic colleague. The former can compare personal experience with its imitation (counterfeit, not facsimile) at every step along its development. Faithfulness to reality guides the programming strategy; it is an unambiguous criterion for choosing techniques. Thus for photorealism, Phong shading and anti-aliasing are invariably superior to mere Gouraud shading and the *jaggies* (aliasing contours), because it makes the image look more like the photograph of the original, the real-thing. On the other hand, consider a rendering of what a four-dimensional artist would have to draw on a three-dimensional canvas in order to evoke the same illusion of an impossible figure as the Penrose tribar popularized by Escher [32, 13, 11]. What does such a 4-D object really and truly look like? Should its image be shaded at all? How much do we care about the jaggies? Actually, we do care. But only after the depiction is totally convincing in its raw-and-ready draft. Only then, resources permitting, might we engage a computer artist to help us to meet the competition for inclusion in it Electronic Theater at SIGGRAPH.[2]

Thus, we find that the metarealist must use other criteria, other standards of comparisons than the photorealist. Among these are fidelity to physical principles, mathematical honesty, even something like a Calvinist economy in programming the RTICA so that its scientific message does not get lost in a baroque programming style. After all, it is usually the student programmer of the RTICA

127

---

[2] *This annual convention of the Association for Computing Machinery draws close to 40,000 visitors and the Electronic Theater presents the industry standard of quality and interest in computer animation. And yet, in 1998 there were only two pieces of thirty with a frankly mathematical content. In 2001 there were none.*

who gains the greatest intellectual benefits from the exercise. Photorealistic perfection intimidates the student programmer and distracts the engaged observer. The latter is expected to think through the mathematical ideas expressed in the RTICA. Thus, the less the artifact resembles something else, something too familiar, the better.

Animating Homotopies

Real-time interactive computer animation is particularly appropriate for illustrating mathematical objects called *homotopies*. These deal with topological changes that happen over time. As such, a homotopy is the natural mathematical abstraction of any temporal change-of-shape. Rather than proposing some grand scheme for animating all possible homotopies, we identify a few *elementary* and *typical* homotopies and develop a vocabulary for exploring their RTICAs.[3] However, we do propose to transfer the notion of homotopy into the vocabulary of computer graphics.

Turning a sphere inside out is the best known of such homotopies [38]. In the four decades since Smale proved its existence, and the two decades since Nelson Max first captured Morin's eversion in a computer animation, there have been many metarealistic renderings of this *Helen of Homotopies*.

Let us review the common meanings of our vocabulary. But, why should we care what the non-specialist, ignorant of the jargon, might hear? Mostly, we do so to anticipate the conscious or subliminal associations our usage will arouse. It also permits us to appreciate the circumstances under which the common term entered the jargon, and the evolution of its meaning along with the technology. A quick dictionary [34] search yields

Animate: Create the illusion of motions.
Motion: Change in position, change of body, gesture, gait.
Animated Cartoon: A motion picture consisting of a photographed series of drawings.
Motion Picture: Series of filmed (photographed) images viewed in sufficiently rapid succession to create the illusion of motion and continuity.

We consolidate these into our working definition of animation:

Animation: A sequence of pictures viewed rapidly enough to evoke an experience of continuous motion.

---

[3] *The students in my courses on geometrical graphics are my chief assistants in this enterprise. I present them a smorgasbord of sample homotopies, often worked on by previous students. They choose one, or come up with a new one, and build a semester project around it. This pedagogical aspect has a healthy influence on my understanding of what is elementary and typical. The website originally created by them, http://new.math.uiuc.edu, continues to present the current and archived work.*

We have less success finding a common word for that which we animate. We borrow the word *homotopy* from topology rather than using the familiar terms: *transformation* and *metamorphosis?* Indeed, both say "change of shape", one in Latin, the other in Greek. But both also have undesirable connotations. In the common tongue, transformation has come to mean just about any kind of change. In mathematics it is synonymous with every function or mapping between two sets. In contrast, the second term has too restricted (entomological and mystical) meanings in the common language. The popular amputation, *morphing* comes closest, but already has a narrower technical meaning. So we borrow a technical term from topology and hope to popularize it.

All uses of the term *homotopy* in topology have this common denominator. Two continuous mappings, f_0, $f_1 : X \rightarrow Y$, from the same source space X into the target space Y are *homotopic* if f_0 and f_1 fit into a continuum of mappings in between. More precisely, there is a continuous mapping

$$h : X \times [0, 1] \rightarrow Y$$

so that for each $x \in X$ we have $h(x, 0) = f_0$ and $h(x, 1) = f_1(x)$. One writes $f_t(x)$ for $h(x, t)$ and refers to f_t as the homotopy from f_0 to f_1.[4] The place that remains the same throughout the homotopy (from Greek: same+place) is X; the *topes* are its continuous images, $F_t = f_t(X)$, depicting what is being deformed in Y. Note that F_t is the deforming shape, and f_t is its parametrization. The neologism *tope* avoids having to import the topological term *image* into a field where image already has too many meanings.

To capture a homotopy into a computer program, both source X^d and target Y^n must be subsets of real Cartesian spaces of dimensions d and n. When $d > n$ we tend to speak of *projecting X to Y*. When $d < n$ we think of *inserting X into Y*. The non-standard term *insertion* collects the notions of *embedding* (globally 1:1), *immersion* (locally 1:1), as well as *Whitney stably singular,* which allows for controlled, non-pathological deviations. When $d = n$, and often as not f_0 is the identity map, then f_1 represents some *rearrangement* of space, and the homotopy f_t is a continuous *distortion* of space until the rearrangement is complete. Think of the commonly used term *warping*, although warping is used more generally for all kinds of bending of objects already inserted in space.

Common examples of insertions ($d < n$) are discrete sets of points ($d = 0$) such as particle systems, curves ($d = 1$), and surfaces ($d = 2$) in space ($n = 3$). To observe the distortion of space ($d = n$) we insert lower dimensional marker objects and follow their fate. We observe projections from space ($d = 3$) to a surface ($n = 2$) either by what appears on the 2-dimensional screen itself, or in terms of pictures affixed to surfaces in space. Ray tracing, texture mapping, and environment mappings should come to mind.

129

---

[4] *Homotopy theory is a major branch of topology today and its concepts permeate many more fields of mathematics, like algebra and analysis. The classic textbook by Seifert and Threlfall [40] speaks of a homotopic deformation from f_0 to f_1. The noun came later.*

In practice, we require the in-between maps, f_t, which parametrize the topes of the animation, to be of the same species as the initial and final maps. Thus, when $f_0, f_1 : X \to Y$ have some structure or regularity, it shall be preserved throughout the homotopy. For example, when X is homeomorphic to each of its topes F_t, the deformation is called an *isotopy*.[5] When $X \subset Y$ and f_t is the restriction to X of a homotopy g_t of Y into itself, we say that it is an *ambient homotopy*. Combining both, an *ambient isotopy* is what we mean by a *distortion* of X. Though it may be difficult to imagine, almost every isotopy we encounter extends to an ambient isotopy.

Classifying Homotopies

We propose a classification of homotopies which takes into consideration the manner in which their RTICA generates the stream of pictures. It is thus hardware and software dependent, but not determined by it. The brief elaboration after each abstract definition will make this clear.

motion: We are moving about a static scene. Or, in the solipsistic coordinate system of our personal reference frame, the entire world is being rigidly moved about in front of us. Either way, we need to understand clearly what *motion* entails.

The customary metaphor of changing *camera coordinates,* of making a *camera path,* is a vestige of the days that an RTICA was used mainly to produce a videotape for a passive audience. Virtual reality and immersive virtual environments are better served by the *observer* metaphor. The world, represented by a *display list* of polygons, X, is subjected to a succession of Euclidean isometries, $M(t)$, before being projected to the screen. In customary graphics programming libraries, such as OpenGL [31], $M(t)$ is a one-parameter family of 4-dimensional matrices applied to the vertices of X expressed in *homogeneous coordinates*. Thus, the abstract group of 3-dimensional Euclidean *isometries* (rotation followed by translation) is represented as a subgroup of the 4-dimensional linear group. Other 3D geometries, such as spherical and hyperbolic, can also be so represented and are treated in a similar way [18]. To the topologist, $M(t)$ is a path in the isometry group of a 3-dimensional geometry. In this context the letter M is a mnemonic for *matrix, motion,* as well as `MODEL_VIEW` in the vocabulary of OpenGL.

articulation: While the scene is not totally static here, and objects move about independently, they retain their geometric shape.

Such a clockwork of rigidly moving components (a marionette) would seem to differ qualitatively from a simple motion. In fact, it does so only quantitatively. One builds a *hierarchy* which assigns to each component its proper motion and

---

[5] *The concept of* homeomorphy *is central in topology; it means "of the same shape." Technically, each f_t is one-to-one and maps X onto its tope F_t in Y; its inverse, $f_t^{-1}: F_t \to X$ is thus well defined and continuous.*

place in the world scene. Hierarchies are also called *scene graphs* because the order in which matrix multiplications are applied to objects forms a directed tree. That is, the i-th object has a display list X_i and an associated isometry, P_i, which places it into the coordinate system of the next higher object.

For Jim Blinn's *Blobby Man* [9], for example, each finger X_i, $i = 1, ..., 5$ would become $P_i X_i$ when it is attached to the hand, X_6, which is attached to the arm X_7, the shoulder, the torso etc. Thus the thumb is rendered as $M_0 ... P_7 P_6 P_1 X_1$, where M_0 incorporates the observers motion. Of course, all the matrices are multiplied together into one before any vertex in the finger is sent down the pipeline. Thus the matrix for the arm is $M_7 = M_0 ... P_7$, and $M_6 = M_7 P_6$ is the matrix for the hand. The world motion, M_0, affects all objects and plays a special role as the inverse of the camera matrix. Suppose we place the observer (ourselves or the camera) somewhere, P_0, in the world. We could build an avatar, X_0, for ourselves so that we appear as $M_0 P_0 X_0$. If now we wish to see through our own eyes then X_0 is rendered without modification, whence $M_0 P_0 = I$ and $M_0 = P_0^{-1}$. If we wished to see through a camera held in the hand, for example, we would replace each M_i by $M_6^{-1} M_i$.

distortion: Here the motion $M(t)$ is not Euclidean, spherical or hyperbolic. We do not assume that the M belongs to any particular group of isometries. But the action MX remains linear (a matrix multiplication). In general, the motion, $M(x, t)$, depends also on the location where it is applied, as well as on time.

A trivial example of this is a change in scale, especially a non-isotropic one, as when a sphere turns into an ellipsoid. Any decoration on the sphere, or an entire articulated hierarchy of objects subordinate to the sphere, would be similarly distorted in the graphics pipeline.

The general case, where M also depends on location, is rarely implemented in a graphics library or in the hardware of a graphics system. We would have liked to use such an acceleration for Lou Kauffman's ambient isotopy of 3-space which illustrates Dirac's String Trick. Here, independent rotations of concentric spheres, parametrized by their radius, have the effect of straightening a twice twisted ribbon [36, 21, 26]. Since it was a distortion of space itself, everything in it (such as very many twisted ribbons) followed along, and no ribbons passed through themselves. Unlike an articulation, a distortion of space guarantees that initially disjoint elements will not collide during the homotopy. As we had no hardware acceleration for this kind of homotopy, it was implemented as a time-consuming per-vertex deformation and the videotape was ultimately made in the traditional frame-by-frame manner rather than capturing a real-time, interactive animation.

deformation: All changes from frame to frame involve changes in the vertices that make up X. Instead of time acting on M, it now acts on $X(t)$. That is, the coordinates of each vertex of an object are changed in time, $x \rightarrow h(x, t)$, before they are sent down the pipeline.

In a sense, all homotopies can be animated as deformations. It is the only way to do it in a graphics system that lacks a hierarchical geometric library with matrix algebra and hardware acceleration. However, when such shortcuts are available, it makes sense to disinguish this residual category from the previous three, which can be more easily programmed and more speedily displayed.

We often combine motions, $M(t)$, with deformations, $X(t)$. But for unfamiliar events it is best to alternate them, drawing $M(\mu(t))X(\xi(t))$, where $d\mu/dt = 0$ whenever $d\xi/dt \neq 0$ and vice-versa. Doing both simultaneously can be distracting unless one *time scale, $\mu(t)$*, is very much slower than the other one, $\xi(t)$.

The Geometry Pipeline

We shall now trace what befalls a display list as it passes down the graphics pipeline by working backward from what you see on the screen. Window systems are ubiquitous, so the absolute *screen coordinates* of a figure have no primary importance. To locate a point in a visible region of the screen (namely in a *window*), we specify fractions of the width and heights taken in the appropriate direction. The x-coordinate, or abcissa, always goes from left to right, but the ordinate is as often top-down as bottom-up. This way the numerical description (of the geometrical objects to be drawn on the screen) is invariant under moving and resizing the window. What to do if the aspect ratio of the resized window has changed, is a matter of choice. Technically, one distinguishes between the window and the *viewport* into which a view of the world is transformed, either orthographically or perspectively. The window then need not coincide with the viewport, especially if they have different aspect ratios. Also, the window may have more than one viewport, as for stereographic pairs.

It is, however, more practical to consider an abstract 3-dimensional space described in *world coordinates,* in which everything happens. On the screen, we see a certain portion of this world through the window. An important motion through this world is a steerable flight through a scene, X. Here X represents all objects already placed into the world coordinates. An incremental adjustment of the controller (mouse or CAVE wand) at moment t is imposed on M_0. We multiply on the left by a small Euclidean isometry $\tilde{M}_t$. Thus $M_0(t + dt)X = \tilde{M}_t M_0(t)X$. The $\tilde{M}$ is a rotation by a tiny angle followed by a translation by a tiny vector. An accumulation of such increments applied to an initial $M_0(0)$ (which is usually the identity) constitutes a flight-path or motion through the world.[6]

Here is what happens to the homogeneous coordinates, (x, y, z, w), of a vertex after it leaves interactive control and passes into the *clipping* part of the geometry pipeline. Only vertices inside the *clipping cube*

$$-1 \leq \frac{x}{w} \leq 1, -1 \leq \frac{y}{w} \leq 1, -1 \leq \frac{z}{w} \leq 1$$

---

[6]   *Unfortunately, OpenGL primitives multiply the current matrix only on the right since that is the natural place for implementing articulated hierarchies. Therefore some matrix arithmetic is inevitable. Moreover, OpenGL departs from the traditional representations of vectors as rows on the left of the matrix to the current standard of column vectors on the right. Even in mathematics this was not always so. Birkhoff and MacLane [8] use the former, and earlier yet, we used Einstein notation, and the issue was moot.*

are admissible. Line segments and triangles that cross its boundary are cropped, $\left(\frac{x}{w}, \frac{y}{w}\right)$ is scaled into the viewport, and $\frac{z}{w}$ is discretized into the depth buffer. The division of (x, y, z) by w is the reason that perspective projections can also be implemented by one last matrix multiplication in the 4-dimensional linear group. The very cleverly designed (but misnamed) *projection matrix* maps a *viewing frustum* homeomorphically into the clipping cube. The viewing frustum consists of a rectangular window and the portion of the rectangular cone from the origin through this window located between the window and a parallel clipping plane further back. One is reminded of Albrecht Dürer's etching of the artist sighting a reclining nude, and marking a transparent canvas where the sight lines pierce it.

We have seen, then, that a vertex in the display list of an object undergoes a sequence of multiplications by matrices representing the Euclidean motions of rotation (about the origin) followed by a translation. More precisely, these modeling matrices are in the 3-dimensional affine group, where the rotation is relaxed to be an arbitrary invertible linear transformation. This allows uniform, and non-uniform scaling. All the power (and frequent confusion) of interpreting the meaning of multiplying a vector by a matrix, and, by extension, the interpretation of the associative law of matrix multiplication, comes into play here. In addition to Birkhoff and MacLane's [8] classical *alibi* and *alias* interpretations, we have a *placement*. An object, described by a display list of coordinates in "its own natural" Cartesian framework, is *placed* into the reference frame of another object higher in the articulation hierarchy.

Let us illustrate the subtle differences in the case of a planar rotation:

$$\begin{bmatrix} u \\ v \end{bmatrix} = \begin{bmatrix} \cos\theta & -\sin\theta \\ \sin\theta & \cos\theta \end{bmatrix} \begin{bmatrix} x \\ y \end{bmatrix} = \begin{bmatrix} x\cos\theta - y\sin\theta \\ y\cos\theta + x\sin\theta \end{bmatrix} = \begin{bmatrix} x\cos\theta - y\sin\theta \\ x\sin\theta + y\cos\theta \end{bmatrix}$$

Customary matrix muliplication "row vector dot column vector" reveals (u, v) to be the coordinates of the same point in a coordinate system which was rotated by an angle of $-\theta$. The point has an *alias*.

$$\textbf{alias} \quad \begin{bmatrix} u \\ v \end{bmatrix} = \begin{bmatrix} \begin{bmatrix} \cos\theta \\ -\sin\theta \end{bmatrix} \cdot \begin{bmatrix} x \\ y \end{bmatrix} \\ \begin{bmatrix} +\sin\theta \\ \cos\theta \end{bmatrix} \cdot \begin{bmatrix} x \\ y \end{bmatrix} \end{bmatrix}$$

Factoring the second expression, we have a trigonometric interpolation which moves the point (x, y) towards its left-perpendicular $(-x, y)$ by an angle of θ along a circular arc. Thus the point has an alibi as to its whereabouts.

$$\textbf{alibi} \quad \begin{bmatrix} u \\ v \end{bmatrix} = \begin{bmatrix} x \\ y \end{bmatrix} \cos\theta + \begin{bmatrix} -y \\ x \end{bmatrix} \sin\theta$$

Rearranging and factoring a different way, shows that (u, v) is what you get if you follow the recipe "go x units along the abcissa followed by y units along the ordinate" when the meaning of abscissa and ordinate has changed by a rotation of the coordinate frame of orthogonal unit vectors along the axes. The point has a new *place* inside a different frame of reference.

$$\textbf{place} \begin{bmatrix} u \\ v \end{bmatrix} = \begin{bmatrix} \cos\theta \\ \sin\theta \end{bmatrix} x + \begin{bmatrix} -\sin\theta \\ \cos\theta \end{bmatrix} y$$

To illustrate the flexibility of our language, let us read

$$P_1\, P_2\, P_3\, X = P_1\, (P_2\, P_3\, X) = (P_1\, P_2\,)(P_3\, X)$$

in two ways. In the first, the object X has been moved about it's own space by the composite motion $P_2\, P_3$ and then placed into the reference frame (coordinate system) P_1. In the second, X has been placed by P_3 into a new reference frame, where it is then moved by the composite motion $P_1\, P_2$.

Permutations of the *alias-alibi-placement* attitudes generate many more verbalizations of varying practicality. However, it is time to apply our new vocabulary to a perennial problem in mathematical visualization, that of seeing into the fourth dimension.

The Fourth Dimension

It is curious how very much the popular mind is intrigued by the *Fourth Dimension* and it would be interesting to speculate why this might be so. See, for example, the article [27] in this volume. But here is not the place for even the most cursory review of this issue. The reader should set aside an evening for browsing the web on the subject. Special attention to the work of Tom Banchoff [4, 6] on the subject will be richly rewarded. What we can do here is to extend our vocabulary for metarealistic rendering a short distance in the direction of the Fourth Dimension.

First of all, we should distinguish between attempts to *visualize* phenomena extended in 4 isotropic dimensions, and our recognizing four and higher dimensional reality by its *special effects* in 3D. Borrowing from media jargon, we might call the former *4D-viz* and the latter *4D-fx*. It is appropriate here to remind the reader that, when all is said and done, we can see only curves and surfaces. We draw inferences about space and bodies by the way curves and surfaces arrange themselves under lighting, occlusion, motion parallax, binocular vision etc. We can learn to draw similar inferences about higher dimensions from the way we furnish ordinary space. We do this in basically four different ways.

decorations: Curves, surfaces and *rooms,*[7] whose points are endowed with more attributes than fit into 3-degrees of freedom, and are equipped with visual artefacts, *glyphs,* which express these attributes.

The whole of scientific visualization might be subsumed under this first rubric. For example, we might paint temperatures on heated objects in vivid color. We distinguish two *hyper-attributes,* say pressure and temperature, by color and texture. Visualizing surfaces in 4-space by painting the fourth dimension on a 3D slice or shadow of it is less successful. Graphs of complex functions are particularly sensitive to such graphical misadventures. The reason is simple. While we are not apt to rotate pressure into temperature, we do want to rotate an object in 4D every which way.

A common way of indicating many hyper-attributes is to place *glyphs* at a sample of points. We remember the little arrows representing velocity vector field from our calculus courses. Glyphs generalize this idea in the form of small, solid shapes, which have an obvious direction. Limited ranges of other attributes are mapped to physical features of the glyph, for example, its size, color, shape, texture, and other details.

charting: Three dimensional subspaces of R^4 are mapped faithfully to a flat, 3-dimensional canvas.

What kind of subspaces, *e.g.* manifolds, and how faithful, *e.g.* conformal, are subjects of a finer classification than we have in mind here. Think of a Mercator or a polar projection of a sphere to a page in an atlas, but both are one dimension higher. Topologists prefer to call it a *chart* rather than a *map* to emphasize that the the mapping is one-to-one and provides a local coordinate system.

Among the most satisfactory examples of this genre are the elegant realizations of non-Euclidean geometries of 3-space in terms of projections from 3-manifolds in R^4. For a positively curved geometry of space we use a *conformal projection* of the 3-sphere from its pole to flat 3-space. The geometers commonly call this projection *stereographic,* which also means *binocular,* and the geographers prefer *polar,* which has yet other geometrical meanings. Straight lines become circles, and geometrically flat Clifford tori become the sensuously oblique Cyclides of Dupin [3]. Central projection of a 3-dimensional paraboloid to the 3-ball gives it a hyperbolic geometry, exquisitely rendered in the the final minutes of the Geometry Center's classic video *Not Knot* [24].

Our real-time interactive CAVE animation, *A Post-Euclidean Walkabout* [18], has four such charts. In the first act we are still in Euclidean space and merely fly around and through a morphing shape of a sea shell. Act two starts with a suitably triangulated and painted hyperbolic octagon in the Poincaré model of the hyper-

---

[7] *The correct technical term is* volumes, *as in* volumetric rendering. *But non-specialists might be misled into thinking of books. The mathematical* sammelbegriff *in question is, of coures,* 3D-manifold. *But one shrinks from the automotive misconstructions of that term. In VR the term* world *is coming into vogue.*

bolic plane. It lifts off the plane and wraps itself into a double torus. Though topologically distorted, the triangulation remains as a testimony to the conformal structure on the Riemann surface. Next, we enter hyperbolic 3-space and fly through its tesselation by right-angled dodecahedra. In the final act we visit the dodecahedral subdivision of the 3-sphere, the 120-cell.

The efficiency of these *4D-fx* is the result of the intrinsically 4-dimensional nature of the Silicon Graphics geometry pipeline, which powers the CAVE. All three isometry groups, Euclidean, hyperbolic and spherical, have a representation in the 4D general linear group which acts on the *projective models* of the three geometries of space. This *Klein Model* of hyperbolic 3-space is *conformal* at the origin.[8] To give the CAVE visitors the correct illusion of flying while keeping them resolutely fixed at the origin, we inflict a 1-parameter family of hyperbolic isometries on the dodecahedral tesselation.

shadows: Generalize to one dimension higher, the familiar perspective, axonometric, and orthographic projections from the fine and graphic arts.

One immediate problem with this is the fact that we look at a perspective drawing of a 3-D scene from the *outside.* Add one dimension and we now experience the picture first hand, by being in it.

However, even with 3D perspective our perception depends on a 2D picture having lots of curves in it: edges, profiles, contours, 1-D textures. Similarly, we can learn to recognize 3D projections of surfaces extended in 4D by the way their shadows on our 3D canvas deform even as the object is rigidly rotating in 4D. This school of thought originated with Tom Banchoff's classic computer animations [5].

Somewhat more difficult to follow is a homotopy of a surface in 4D, such as the unraveling of an unknot. An embedding of a 2-sphere in 4D is called a *4D-knot.* It is a trivial 4D knot, an *unknot,* precisely if an isotopy moves it into a 3D-flat, and there, it looks like a sphere. The lower dimensional analogue of such a thing is an unknotted tangle, say your garden hose with end screwed to end, magically untangling itself, without *cutting and pasting,* so as to lie flat on the lawn as a perfect circle. This was Dennis Roseman's subject in our CAVE tryptich, *Laterna matheMagica* [19]. There, the *shadow* of the unknot was decorated with a color to indicate the intersection with a 3D-flat slicing through the surface.

slices: If we interpret one axis in R^4 as time, then the succession of orthogonal 3-spaces organize themselves into one spatio-temporal experience, *e.g.* a movie, an earthquake, a dream. Conversely, any process in space-time can be regarded as a monolithic 4-dimensional entity.

---

[8]   *Confomal means angle-preserving in general. The "Poincaré Model" of hyperbolic space is conformal at all of its points. Its straight lines are Euclidean circles. To be conformal just at the origin means that visual angles from that viewpoint are the same in hyperbolic geometry as in the Euclidean geometry in which the model is constructed. But that suffices for distant angles to also look correct to our Euclidean eyes. In particular, right angles look right.*

If we slice a plane through a surface in 3D we see curves wriggling in the plane. The curves need not be amorphous. Floor plans of buildings are also 2D slices of surfaces in 3D, which change discontinuously as we move through the ceiling. Contour lines of maps depicting mountain tops, passes and valleys are familiar examples of slices. If we pass a 3D slice through a surface in 4D we get curves in space which undergo interesting recombinations. Of course, we might start with the recombinations and organize them into a coherent surface in 4-space.

However a better analogy is obtained by considering a maze. A 2D-maze is what we solve on a piece of paper. Even when we are in the maze, as in a park or on a floor of an unfamiliar building, we can find our way out as soon as we see a floor-plan. A 3D-maze is just a stack of floor-plans, a building for example. So a 4D-maze is a stack of buildings, and finding your way around and out of such a maze is best left to Virtual Reality [35].

Technology-driven Research

From the beginning of my topological apprenticeship I was repelled by the customary highly formal, mostly verbal, and poorly illustrated descriptions of homotopies. In time, I also learned that the authors of these descriptions secretly drew pictures for themselves to clarify their thoughts, to simplify, and even to discover new theorems. Only their modest skills and lack of training in the graphic arts kept them from making their expositions easier to follow. Or so I thought! In fact, it was the custom in the heyday of Bourbaki to disparage pictures in mathematics. It was rationalized by claims that pictures lead to erroneous inferences, that they are expensive to print, that they take up space better devoted to elaborate notation.

Determined to illustrate homotopies more effectively, and to teach others to do likewise, I needed to find new graphic media. For my purposes, the medium had to be completely under my control. Inspired by Hilbert CohnVossen [30], I first learned to draw with pencil, chalk and ink pen, on paper napkins, blackboards and drafting vellum [20]. But some homotopies simply would not fit into a set of one, two, or however many discrete pictures; they required animation [25].

Traditional cell-animation (á la Disney) was out of the question. This medium requires substantial group effort. Cell-animation teams are highly differentiated and specialized. Having little talent for management and no appreciable resources, I had to look for other ways to animate my homotopies. Computer animation looked attractive, in particular the promise of real-time interactive computer animation. Once early on, I was too late to make a videotape in time for a public presentation. The convention then (and still today) was to capture frames into animation buffers, transfer them over the the the net, and lay them onto the tape. My NCSA colleague, Ray Idaszak, suggested that I build flexible steering gadgetry into the program and wire the *Personal Iris* workstations directly to the beta-cam recorder. I rehearsed my talk as I steered the RTICA, and the tape got done in time for the conference.

Computer animation, which meets current standards of quality, continues to be produced in the traditional stop-action manner, because computers cannot produce frames of the required quality in the twenty-fourth of a second needed for the animation to pass the threshold of fluidity. As computer speed catches up to yesterday's standards, tomorrow's techniques of photorealism streak out of range of real-time interaction. Typically, an RTICA is used for a computer animation only initially: to explore and refine, perhaps to choreograph the action and plan the camera paths. At this stage of the production, rough, sketch-like rendering suffices. Later, often at tremendous cost in resources (time, cycles and storage) the planned frames are rendered and stored at leisure until they can be assembled into a videotape or digital movie. By and large, this process was not for me.

Some of my projects did require division of labor and specialization to be *successful,* to result in SIGGRAPH quality videotapes produced in the traditional manner [10, 22, 36, 39]. These projects required far too much emotional effort and their completion generally left me drained and unproductive for months! Quantitatively speaking, it was more cost-effective to concentrate on the construction of the RTICA than on the production of competitive videotapes.

The decision to optimise the efficiency of an RTICA, and to find homotopies most suitable for the technology at hand, eventually paid off handsomely. Realtime interactive animation was exactly what was needed for immersive virtual environments, such as the CAVE [15, 17, 18, 19, 37]. This new medium presented its own geometrical and pedagogical challenges. Adapting CAVE programming to the classroom involves careful design of examples and prototypes like the *illiShell* (1994) and *illiSkeleton* (1998).

My *illiView* project[9] acquired new shibboleths like "transfer of technology" and "rapid prototyping". Each new technological opportunity and demand generated criteria for choosing the homotopy to be taken on next. Not infrequently, a student in a geometrical graphics course would express the desire to experiment with a particular set of graphical features unfamiliar to me. In keeping with the mathematical nature of the courses, this also required finding a suitable homotopy to be illustrated with the new technology.

This, then, is the context in which our present article takes its origins. Our modest expansion of the technical vocabulary for treating real-time interactive computer animation with some precision, at least fits our experience. We hope that it will prove equally useful to other graphicist who practice metarealistic rendering, whether they approve of this name for their work or not.

---

[9] *Named in admiration of the Geometry Center's geometrical viewing package,* Minneview, *its highly successful successor,* Geomview, *and cousin* Meshview *by Andy Hanson,* illiView *differs from these in many respects, as described in [25].*

Bibliography

[1] *Proceedings of the First International Symposium on Non-Photorealistic Animation and Rendering.* ACM-SIGGRAPH, Eurographics, Annecy, France, June 5-7,2000.

[2] Hans-Dieter Abring. *Von Daguerre Bis Heute, Band I.* Buchverlag und Foto-Museum, 1990.

[3] Thomas Banchoff. *The spherical two piece property and tight surfaces in spheres.* Journal of DifferentialGeometry, 4(3):193–205, 1970.

[4] Thomas Banchoff. *Beyond the Third Dimension.* W. H. Freeman&Co., Scientific American Library, New York, 1990.

[5] Thomas Banchoff and Charles Strauss. *Complex Function Graphs, Dupin Cyclides, Gauss Maps and the Veronese Surface.* Computer Geometry Films: Brown University, Providence, RI, 1977.

[6] Tom Banchoff and Davide Cervone. Math awareness month 2000: An interactive experience. In this volume p. 83–97.

[7] David Banks. *Interactive display and manipulation of two dimensional surfaces in four-dimensional space.* In *Symposium on Interactive 3D Graphics,* pages 197–207. ACM, 1992.

[8] Garrett Birkhoff and Saunders MacLane. *A Survey of Modern Algebra.* MacMillan, 1953.

[9] Jim Blinn. *Jim Blinn's Corner: A Trip Down the Graphics Pipeline.* Morgan Kaufmann, 1996.

[10] Donna Cox, George Francis, and Ray Idaszak. *The Etruscan Venus.* In Tom DeFanti and Maxine Brown, editors, *Video Review Number 49.* ACM-SIGGRAPH, 1989.

[11] H.M.S. Coxeter, Michele Emmer, Roger Penrose, and M. Teuber, editors. *M.C. Escher: Mathematics and Art.* North Holland, Amsterdam, 1986, 5th edition.

[12] Carolina Cruz-Neira, Daniel J. Sandin, Thomas A. DeFanti, R. V. Kenyon, and John C. Hart. *The CAVE: Audio-visual experience automatic virtual environment.* Communications ACM, 35(6):65–72, 1992.

[13] Michael Daniels and Michael Pelsmajer. *Illusion.* National Center for Supercomputing Applications, University of Illinois at Urbana-Champaign, 2000. An *illiView* real-time interactive CAVE animation.

[14] Michele Emmer and M. Manaresi, editors. *Mathematics, Art, Technology and Cinema.* Springer Verlag, New York, 2003.

[15] John Estabrook, Ulises Cervantes-Pimentel, Birgit Bluemer, and George Francis. *Partnerball: a mathematical experiment in simulating gravitational lenses in tele-immersive virtual reality.* Supercomputing98, 1998.

[16] Mark Flider. *Schprel.* National Center for Supercomputing Applications, University of Illinois at Urbana-Champaign, 2001. An *illiView* real-time interactive CAVE animation.

[17] Alex Francis, Umesh Thakkar, Kevin Vlack, and George Francis. *CAVE Gladiator: Experiments on the relative value of head-tracking and stereopsis in CAVE immersive virtual environments.* CAVERNUS: The CAVE Research Network Users Society, 1998.

[18] George Francis, Chris Hartman, Glenn Chappell, Ulrike Axen, Paul McCreary, Joanna Mason, and Alma Arias. Post-Euclian Walkabout. In *VROOM – the Virtual Reality Room.* ACM-SIGGRAPH, Orlando, 1994.

[19] George Francis, John M. Sullivan, Ken Brakke, Dennis Roseman, Rob Kusner, Alex Bourd, Chris Hartman, Ulrike Axen, Jason Rubenstein, Paul McCreary, Will Scullin, and Glenn Chappell. *LATERNA matheMAGICA.* In Holly Korab and Maxine D. Brown, editors, *Virtual Environments and Distributed Computing at SC'95: GII Testbed and HPC Challenge Applications on the I-WAY.* ACM/IEEE Supercomputing'95, San Diego 1995.

[20] George K. Francis. *A Topological Picturebook.* Springer, 1987.

[21] George K. Francis and Louis H. Kauffman. *Air on the Dirac Strings.* In William Abikoff, Joan Birman, and Katherine Kuiken, editors, *The Mathematical Legacy of Wilhelm Magnus,* volume 169, pages 261–276. Amer. Math. Soc., Providence, RI, 1994.

[22] George K. Francis, John M. Sullivan, Stuart S. Levy, Camille Goudeseune, and Chris Hartman. *The Optiverse.* In *SIGGRAPH Video Review,* volume 125 of *Electronic Theater.* ACM-SIGGRAPH, 1998. 3 min narrated and scored videotape.

[23] Stuart Green, David Salesin, Simon Schofield, Aaron Hertzmann, Peter Litwinowicz, Amy Gooch, Cassidy Curtis, and Bruce Gooch. *Non-Photorealistic Rendering, Course Notes.* ACM-SIGGRAPH, New Orleans, 1999.

[24] Charlie Gunn and Delle Maxwell. *Not Knot.* A K Peters, Wellesley, MA, 1991. video (17 min) produced by the Geometry Center, U. Minnesota.

[25] Andrew Hanson, Tamara Munzner, and George Francis. *Interactive methods for visualizable geometry.* IEEE Computer, 27(4):73–83, July 1994.

[26] John C. Hart, George K. Francis, and Louis H. Kauffman. *Visualizing quaternion rotation.* ACM Transactions on Graphics, 13(3):256–276, 1994.

[27] Linda Henderson. *Four-dimensional space or space-time: the emergence of the cubism-relativity myth in new york in the 1940s.* M. Emmer, ed., *Visual Mind 2.* MIT Press, Cambridge, MA, 2005.

[28] Linda Dalrymple Henderson. *The Fourth Dimension and Non-Euclidean Geometry in Modern Art.* Princeton University Press, Princeton, new edition MIT Press, Cambridge, 2001 edition, 1983.

[29] Aaron Hertzmann and Denis Zorin. *Illustrating smooth surfaces.* SIGGRAPH2000 Conference Proceedings, pages 517–526, 2000.

[30] David Hilbert and S. Cohn-Vossen. *Geometry and the Imagination.* Chelsea, 1952.

[31] Mark J. Kilgard. *OpenGL programming for the X Window System.* Addison-Wesley, 1996.

[32] Scott E. Kim. *An impossible four-dimensional illusion.* In David W. Brisson, editor, *Hypergraphics: Visualizing Complex Relationships in Art, Science and Technology,* pages 187–239. Westview Press, Boulder, 1978.

[33] Lee Markosian, Michael Kowalski, Samuel Trychin, Lubomir Bourdev, Daniel Goldstein, and John F. Hughes. *Real-time nonphotorealistic rendering.* Proceedings of SIGGRAPH 97, pages 415–420, August 1997.

[34] William Morris, editor. *The American Heritage Dictionary of the English Language.* Houghton Miffin, 1969.

[35] Michael Pelsmajer. *4D-Maze.* National Center for Supercomputing Applications, University of Illinois at Urbana-Champaign, 1999. An *illiView* real-time interactive CAVE animation.

[36] Daniel Sandin, George Francis, Louis Kauffman, Chris Hartman, Glenn Chappell, and John Hart. *Air on the Dirac Strings.* In Tom DeFanti and Maxine Brown, editors, *Electronic Theater.* ACM-SIGGRAPH, 1993.

[37] Daniel Sandin, Louis Kauffman, George Francis, Joanna Mason, and Milana Huang. *Getting Physical in Four Dimensions.* In *VROOM – the Virtual Reality Room.* ACM-SIGGRAPH, Orlando, 1994.

[38] John M. Sullivan. *The Optiverse and other sphere eversions.* In *ISAMA 99,* pages 491–497. The International Society of The Arts, Mathematics and Architecture, Univ. of the Basque Country, 1999. Proceedings of the June 1999 conferece in San Sebastián.

[39] John M. Sullivan, George Francis Stuart S. Levy, Camille Goudeseune, and Chris Hartman. *The Optiverse.* In Hans-Christian Hege and Konrad Polthier, editors, *VideoMath Review Cassette,* VideoMath Festival at ICM'98. Springer-Verlag, Berlin, 1998. 6.5 min narrated and scored videotape.

[40] H. Seifert und W. Threlfall. *Lehrbuch der Topologie.* Chelsea, 1947.

[41] *Webster's Collegiate Dictionary.* Thomas Allen, Ltd., 5th edition edition, 1936.

[42] Jeffrey R. Weeks. *The Shape of Space.* Dekker, 1985.

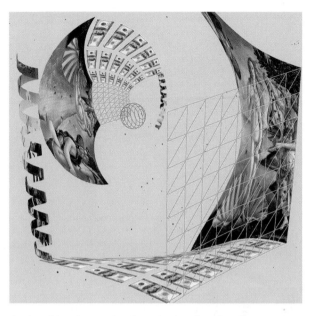

Fig. 1. Snapshot into a gravitational lensing project by John Estabrook, Ulises Cervantes-Pimentel, Birgit Bluemer and George Francis. In this real-time interactive CAVE animation an invisible mass distorts our view of the world. A second image forms within the Einstein radius about the mass, which is inside the spiral ball. The ball is confined to bounce about the cubical stage. The rules of this CAVE-to-CAVE game, based on the Prisoners Dilemma and implemented on the DuoDesk, amused the NCSA PACI-Partners and visitors to Supercomputing98 [15]

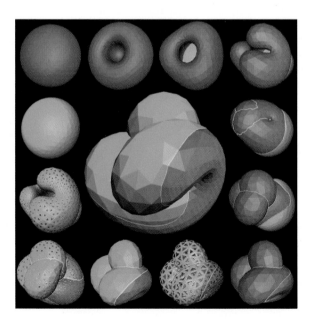

Fig. 2. Thirteen topes (stages) of the symmetry-3 eversion in the "Optiverse" by John M. Sullivan, Stuart Levy and George Francis [19, 22, 39]. Clockwise (from top-left), this regular homotopy turns a bi-colored sphere inside out by passing through Boys Surface (center). It is not an isotopy of the sphere in 3-space, nor can it be regarded as the shadow of an isotopy in 4-space. In metarealistic terms this famous eversion is an essential deformation

Fig. 3. Thirteen topes (stages) of the "illiSnail" animation in the "Post-Euclidean Walkabout" CAVE show at SIGGRAPH94 by Chris Hartman, Glenn Chappell, Ulrike Axen, Paul McCreary and George Francis [18]. Clockwise (from top-left), we chart Blaine Lawsons ruled, minimal surfaces in the 3-sphere by projecting them conformally to 3space, passing through a meridian 2-sphere (1), a half-twist Möbius band (2) with a circular border (3) closing up to Steiners cross-cap (4) and Roman surface (5). A once-twisted ribbon (6,7) closes up to the Clifford torus, seen from the outside (8) and inside (center). A 3-half twisted ribbon (9) closes up (10) to half of Lawson's minimal Kleinbottle. This surface is also the mapping cylinder of $w^2 = z^3$, and Ulrich Brehm's trefoil knot-box. This real-time interactive CAVE animation has conformal projections (charts), shadows and slices of surfaces embedded, rotating, and isotopically deforming in 4-space

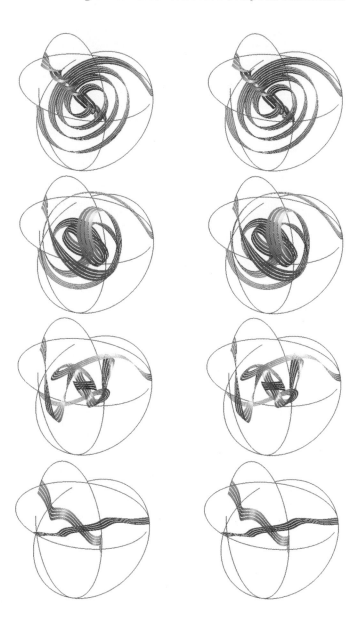

143

Fig. 4. Four cross-eyed stereograms from the "Air on the Stringsof Dirac" by Dan Sandin, Lou Kauffman, Chris Hartman, Glenn Chappell, John Hart and George Francis [36, 21, 26] presented in the Electronic Theater, SIGGRAPH93. Any number (here 4) of ribbons, each with two full twists, kept stationary at the center and the periphery of the orb, untwist in the space between without gettting tangled up. This ambient isotopy of space is an example of a distortion which is a special effect of the quaternionic geometry of the group of Euclidean rotations

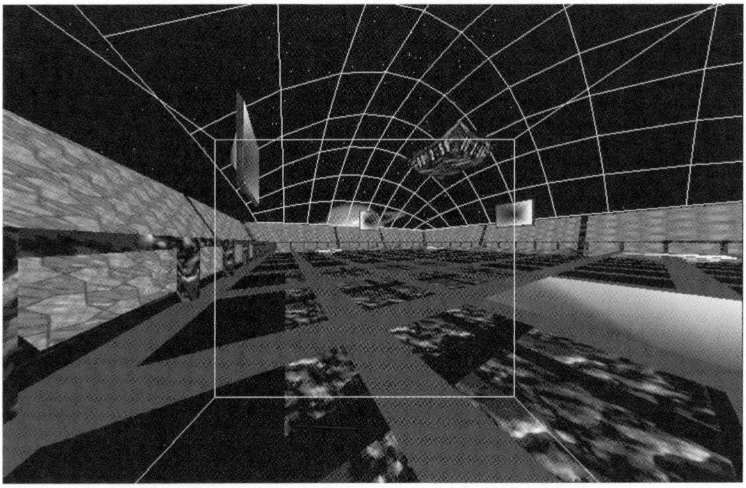

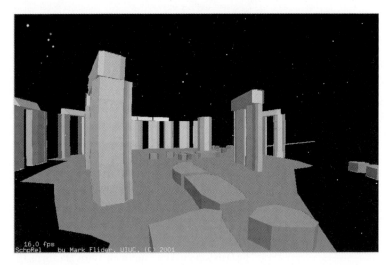

Fig. 5. The arena for the physics based game, "CAVE Gladiator", by Kevin Vlack, Alex Bourd, Alex Francis, Umesh Thakkar and George Francis [17]. This fanciful and wholly non-violent game, combining elements from ice hockey, basketball, and archery, was used for a human factors experiment. Binocular vision is obviously essential in the 5 foot radius near-field, and irrelevant in the vista-space at "infinity". We confirmed the conjecture that, in the arena-size action space (5 to 100 feet), binocular vision is less important than other depth-cues such as motion parallax, occlusion, and perspective

Fig. 6. Stonehenge scene from Mark Flider's special relativity CAVE animation, "Schprel" [16]. To simulate the experience of an observer steering her way through the landscape, ideally, every point at each instant is displaced depending non-linearly on the position and velocity of the observer now, and for all time in the past. For an approximation of this illusion only a few past positions of each vertex are cached. Then, the next position of the observer elicits a computation of the historical place of each vertex whose light reaches the observer now

Virtual Sculptures

Herbert W. Franke

The method of computer graphics makes it possible to visualize ideas for sculptures before its production with material. Here are to consider the physical and technical conditions for the realization, for instance for the stability. Used in this way the method brings advantages for planning, preparing and describing the shape, and especially also for the production: It is possible to transform the program code so that an application for computer controlled milling and cutting machines is possible.

This is a new method for realize art works, it is possible to produce the sculptures very rapidly with high precision, but the results correspond to the classical type of sculpture. Otherwise the most interesting purpose by using the computer as a tool of art in general is the possibility to enlarge the repertoire of art works. The question is: Could the computer lead us also in new areas of sculptures? A way to find an answer is to ignore the mentioned conditions for physical realization and try to design not realizable 3D-forms. It is remarkable that the computer graphic systems allow us to generate pictures also of such objects.

What can be the reason that an object could not practically realized? Here I give a little list of items:

- mechanical instability,
- ignorance of the gravitation,
- irrealistic movements,
- change of shapes,
- permeations,
- not realizable effects of light and colors
- and so on

A good way to illustrate these ideas is to show some examples.

Remark: The program code is written in *Mathematica* 4.1. With *Mathematica* versions higher than 3.0 it is possible to calculate the series of pictures – for this reason eliminate the line *"DisplayFunction -> Identity"*. For better resolution decrease the iteration steps. You can download it from my website – the address is: http://www.zi.biologie.uni-muenchen.de/~franke/VirtS1.html.

Examples

1. Spindle

The basis of this example is a helical twisted spindle – an interesting object because of his optical irritating qualities: Showed in rotation it gives the impression of a screw movement up or down.

I show a superposition of two spindles, rotating in opposite directions. This leads to an object with selfpermeation, principally not to realize.

Fig. 1. Illustration of example "spindle"

2. Snail

This example concerns to an object on the basis of the ENNEPER minimal surface. His form is changed by introduction of further terms into the formula. The movement is a combination of rotation und variation with help of changing variable parameters in the introduced terms. Also here it is to constate a selfpermeation with wandering cutting lines.

Fig. 2. Illustration of example "snail"

3. Donut

The principle of this object is morphing between three mathematical forms of surfaces, the first based on a sphere and the second and third based on a transformed toroid. The result is an unusual movement of pulsation and torsion.

147

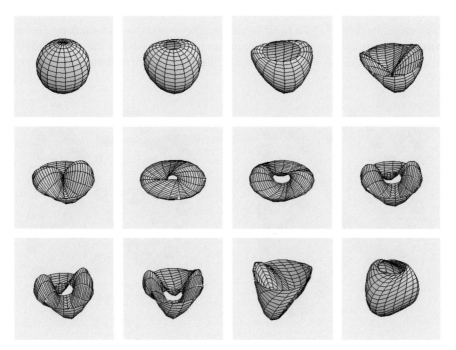

Fig. 3. Illustration of example "donut"

4. Wheel

The object is build up with two parts. The extension is changing during the rotation and also the colors depend on the angle.

Fig. 4. Illustration of example "wheel"

5. Chain

The object in the form of a chain is put together with little dodecaeders. Two of such chains are moving so that one sequence of cubes grasp into the other without touch. The 3-dimensional process should be presented in stereo view.

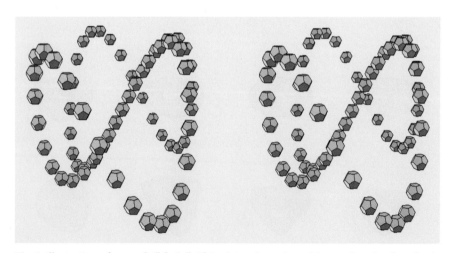

Fig. 5. Illustration of example "chain". This picture is to view with crossing visual angles (a little glass of whisky could be helpful!).

Summary

The results of this experiments have a strange position between reality and irreality. In spite of the impossibility of the physical realisation they have a concrete 3-dimensional form. The medium for presentation is the animation on the monitor or on the television screen. Also a 3-dimensional presentation is possible with help of stereo view or also with a holographic output device. Because the virtual reality will become increasing meaning in the art of tomorrow, the question of construction with material will become negligible. Cyberspace sceneries are good places to use it, but they could also be used in web design – as moving emblemes or simply as eye catcher. And also in internet galleries the possibility of physical production is not weigthy.

The virtual sculpture, as discussed in this contribution, could lead far away from the classical type of sculpture in direction to abstract films or to the mobile art objects. But, and this is my opinion, it should - after all the transformations and movements – remain an uniform object, for instance based on the conception of cyclic processes or on random controlled deviations from the prototype.

In the art the theoretical discussions should not have decisive meaning for the evaluation, the question, if the results are innovative and aesthetically relevant is more important. Without doubt the ignorance of physical and technical conditions opens a door to never seen objects and movements, and the occupation with this is stimulating for the creator and interesting for the spectator – especially as consequence of confrontation with new spatial structures. In the large field of computer art the virtual sculpture will be only a little facette, but here can originate fascinating results.

149

Bibliography

[1] Herbert W. Franke: *Animation with Mathematica,* Springer-Verlag, Heidelberg, 2001
[2] Stephen Wolfram: *The Mathematica Book,* Wolfram Media, Champaign, Il., and Cambridge University Press, Cambridge, UK, 1999

Publication of Electronic Geometry Models[1]

Michael Joswig and Konrad Polthier

Introduction

We have set up a new journal for electronic geometry models at the website http://www.eg-models.de/. The collection exhibits peer-refereed data sets of geometry models from a broad range of mathematical subjects including, but not restricted to, differential geometry, discrete geometry, computational geometry, topology, and numerical mathematics. The published models are intended to provide insight into geometric shapes, to serve as a unique reference for scientific datasets, and to enable the validation of numerical experiments. We do have images and we do have interactive means for visualization. But, it is essential that the focus of this server is not restricted to the visualization aspects. The key item is the data set itself combined with a self-contained description which states why the model is mathematically important [8] [7].

Mathematical models have a long history even if they have been in the background for a longer period. Plaster models were heavily used in the 19th century for educational purposes and, moreover, mathematical research, see Figure 1. For example, the collection of mathematical models in Göttingen already had a long history when Felix Klein and Hermann Amandus Schwarz took over the direction and systematically completed the collection for education in geometry and geodesy. The publisher Martin Schilling in Halle a.S. [14] was specialized on the manufacturing of mathematical plaster models.

The idea of this electronic model server is to continue the plaster collections with modern computer tools. But the possibilities of the digital models go well beyond those of the libraries with classical plaster shapes and dynamic steel models in earlier days. For one thing the digital models are accessible through web browsers and accessible world-wide. Very many of the models are displayed in a fully interactive viewer. Users can adjust the camera and viewing options, for instance, to simulate a walk through a model.

Much more rewarding, however, is the option to use the data sets on the EG-Models server for one's own computations and experiments. Many interesting

[1] An extended version of this article is published in the book: *Multimedia Tools for Communicating Mathematics*, J. Borwein, M. Morales, K. Polthier, J.F. Rodrigues (Eds.) Springer Verlag 2002.

geometric objects are generated by computer programs today, the reason being that they are too large or too complicated to construct by hand. The Internet is full of interesting collections, but often the data made available is in some obscure format or it lacks information, which makes it hard to use. For our models we have a set of non-proprietary ASCII based standard file formats, which are documented on the server. Many software tools can use these formats right away, and it should even be easy to make one's own software read any of them. Besides this, nonetheless important, technical detail, there is a much more severe problem: the integrity of the data provided. This is why it is crucial to run such a server for digital models like a refereed scientific journal.

The collected models are from two different categories. First, the archive is open for electronic counterparts of classical known geometries for educational purposes and for reference. Second and most importantly, the server hosts new models and previously unknown results of mathematical experiments whose existence contributes to mathematical knowledge. Models of the second kind are peer reviewed before publication like an article contribution for a mathematical journal. After acceptance and publication these models will further be reviewed by Zentralblatt für Mathematik. The access to the model server is free of charge.

There is a wide range of possible applications for our kind of electronic models: education and analysis of shapes, comparability and verification of experiments, as well as models which further research can be founded on. The educational aspect was one of the driving forces behind the historic plaster collections. This continues to be true for electronic models. Users may interactively study complex shapes similar to the study of physical geometries. Even more with a digital model, users may interactively study complex shapes much as they study physical geometries. The set of tools operating on normalized shapes will be continuously growing.

Computational experiments based on numerical algorithms are often hard to compare due to individual details in the modeling. In this context it is a major advantage to have a unique model to base further computations upon. Several different research groups may solve the same problem with different algorithms or software packages. Their respective performance can be compared on standard models. Such a comparison of mathematical experiments is currently not possible since classic journals usually do not allow publishing experiments which include complete digital data yet. At most, a description of the experiment contains a few sample figures, which are often useless for other experimenters. One reason is the lack of space in printed journals. But, even with the appearance of electronic journals the classical style of publication of experiments did not change by much. The EG-Models server not only provides a place for the publication of experiments but it also specifies formal criteria for the data sets to be published. As a further benefit standardized models establish perfect test cases for new algorithms and implementations.

The server grew out of discussions at the conference on the "Future of Mathematical Communication" (FMC) in December 1999 at MSRI, see http://www.msri.org/calendar/workshops/9900/. At that time, the software tool JavaView [11] was already developed for doing interactive geometry on the Web and had proven its usefulness in a sample application for the "Dissertation Online" project of the ma-

 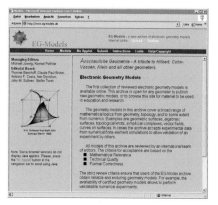

Fig. 1. Plaster model of the Kuen surface (left) and homepage of the new EG–Models server (right).

jor German academic libraries. Several participants of the FMC conference suggested using JavaView to setup an electronic model server for certified and reviewed geometry on the web. During the beginning of the year 2000, we sketched an outline of the server functionality as well as of the software tools required. To establish specifications of the underlying document data formats turned out to be the most crucial step. All documents for the textual model description as well as the model data format are now specified in the new XML language, see Section 4.

153

Publishing Models in an Electronic Journal

There are at least three important reasons for the publication of refereed geometry models:

1. Validation of experimental data sets
2. New research based on refereed geometry models
3. Establishing intellectual property rights

The publication of mathematical experiments is nowadays widely accepted as a source of inspiration and a test bed for theoretical ideas. Mathematical experiments seem to get an important role similar to experiments in physics. There already exist mathematical journals with a special focus on the publication of experimentally obtained results. For example, *Experimental Mathematics* (http://www.expmath.org) and electronic journals like *The Electronic Journal of Combinatorics* (http://www.combinatorics.org), *Documenta Mathematica* (http://www.mathematik.uni-bielefeld.de/documenta/) or *Geometry and Topology* (http://www.maths.warwick.ac.uk/gt/gtp.html) already have a first class status. The dramatic influence on mathematical knowledge acquisistion may be best seen on the effect of preprint servers like the server in Los Alamos (http://xxx.lanl.gov) and its multiple mirror sites. Nevertheless, publications on experiments still remain textual descriptions which hardly allow other researchers to validate the experiments using their own software tools, or to use the experimental data as a basis for own experiments.

A Model Submission

A model is represented by a master file which is a unique and well-defined digital data set of the geometric shape or of the experimental result. This master file is the reference data set of the geometry model, it is the central component of the publication. A model may be generated using an arbitrary software tool, but it must be stored in one of a few geometry file formats. A valid submission of a model consists of the following set of files:

- A data file of the model, the so-called master model. Often this file consists of a polyhedral mesh of the shape of the geometry including further information.
- A description file of the model with a self-contained explanation of the model, and author, content and bibliography.
- Optionally, an unrestricted set of additional files for various purposes. For example, a preview image, a Postscript image for inclusion in paper based articles, files for explanatory purposes, or files usable in other software packages.

For example, the submission of the deformation retract soap film [1] by Ken Brakke consists of the set of files displayed in Table 1.

These files must be produced by an author with his or her private tools or with any software product. The files are then uploaded by the author using the submission page of the EG-Models archive. The upload initiates an automatic verification of the uploaded description and geometry files. The master model specifies the shape of the geometry and is the most important data set of each submission. It is important for comparability of models and the verification of experiments that the shape of each model is uniquely defined by its data file.

The quality of a model depends on the syntax of the data file and on the semantics of the mesh of the model. The specification of the data file must allow a verification of the content of a given data file against a language specification of the syntax and it must allow an unambiguous reconstruction of the model. We have

Table 1: Files in the submission of the model [1]

Retract.xml	The description of the model including author information
Retract_Master.jvx	The master data set of the model
Retract_Applet.jvx	A small version for fast preview in an interactive applet
Retract_Preview.gif	An image for fast preview on the web site.
Retract_Print.ps	PostScript image for inclusion in TeX publications
Retract.fe	Private file format for software Surface Evolver

chosen an XML syntax for the data file which allows an automatic validation of a given data file against the corresponding DTD.

Description of a Model

It is essential that a model submission includes a self-contained detailed description file explaining the mathematical relevance of the model and often details of the experimental setup. The format of a description file must follow exact rules to enable mechanical processing, for example, an automatic generation of a web presentation or print version.

We decided to base most of the EG-Models documents on the XML data format. XML is designed for the automatic validation and the mechanical processing of documents. In particular, XML documents are easily converted into different display formats like, for example, an HTML web page.

Note that an author neither needs to know about XML nor to type any XML by hand. There are different ways to generate model description files, for example, to fill out a template file using the online submission form on the EG-Models server.

Part of the reason why the plaster and string models from the nineteenth century ceased to be effective as teaching and learning tools is that the accompanying material was usually not displayed, and often it was lost, so that most viewers had no idea what the model stood for. Fischer's [2] two-volume set "Mathematische Modelle" helped to rectify that situation, and updated things at the same time. It seems that we want the new electronic model collection to be at least as useful as those volumes, and that means that we have to provide access to background and supporting materials. This does not have to add a good deal of material to the first description people see, but it does suggest that there can be a reference to a more complete treatment of the model. Since we have a list of models, we should also have a list of additional pages that go with some of the models, and perhaps in some of these pages, we can even give the information someone would need in order to make a model, concrete or electronic, static or dynamic.

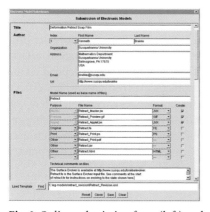 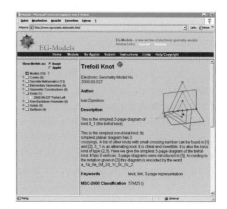

Fig. 2. Online submission form (left) and a published model (right).

Refereeing Process

Each submission is peer-refereed by at least one editor or an external referee. Based on this review, the formal correctness, mathematical relevance, and technical quality the editorial team decides about acceptance of a model. Such strict review criteria ensure that users of the EG-Model archive obtain reliable and persistent geometry models.

Formal Correctness

Each model submission must consist of a master file containing the geometric data of the model itself as well as a description, see Section 2. All the files comprising a submission are strictly required to satisfy a variety of formal criteria. This is to allow the file handling on the server done automatically – at least to a large extent.

The master geometry file must fit the format specifications of any of the geometry file formats JVX, OBJ, POLY, BYU. The server contains a full description of these formats at http://www.eg-models.de/formats/. These standard file formats are easy to convert into other file formats.

The description file must be an XML file which validates against the data type dictionary (DTD) at http://www.eg-models.de/rsrc/eg-model.dtd. This DTD is the formal specification of the XML description file, see Section 4. It is recommended to supply a (GIF or JPEG) image for previewing and a suitably small (JVX) data set for interactive visualization on the Web. Additional files for information purposes are welcome.

Mathematical Relevance

Each model must stand in relationship to a distinguished problem of mathematical importance. Typical examples are the following.

- The model yields a counter example to a previously open conjecture.
- The model provides the first numerical solution of a problem.
- The model is another solution to an already solved problem with distinguished new properties.

Technical Quality

We want to exhibit models which are as useful as possible. Especially for later processing with other software tools, we require the models to satisfy a range of criteria in addition to what can be formally specified. The individual criteria depend on the area the model comes from. For example, for topological investigations it is crucial to have a combinatorial description of a mesh which encodes a triangulation or cell decomposition of the surface. There exist several software tools for optimizing geometries. For example, JavaView allows simplifying meshes of geometric surfaces under curvature control. The technical quality of numerically obtained geometry models is hard to measure, and the editors will decide case by case.

XML for Descriptions and Models

XML is the successor of HTML and is becoming the new lingua franca of the internet. The XML language plays a role at several stages on the model server. Most importantly, it is used for the textual model descriptions as well as for the geometry files (if the JVX format is used). Internally, it is also used within our refereeing system and for other "housekeeping" purposes. The two main advantages of XML are:

1. Straightforward convertability of the data into various other (including future) formats.
2. Syntax check of the data (validation).

This first property allows displaying the content of the server in different views. For example, a referee of a model sees more than the user of the server. XML provides means to produce views of one model in all kinds of abstractions. Further, the XML format allows an almost automatic conversion of the model data into other formats. HTML, as used on the server, is only one example. The JVX files can easily be converted in standard formats used by other programs such as BYU, OBJ, etc. For a project like the EG-model server it is essential to take into account that none of the currently used file formats will survive forever. XML has the capability to cope with exactly this problem.

There are two key properties of XML which are perfectly expressed in its name. Firstly, XML is a Markup Language, that is, XML imposes a structure on the data stored. Here the structure essentially is a rooted tree with arbitrarily many children for each node. This tree structure is the key for processing the data in a standardized way. In contrast to HTML, where the structuring tags are mainly used for specifying layout datails, the XML document itself typically stays completely neutral with respect to layout. The second crucial thing about XML is its extensibility. Again a comparison to HTML might help. While the latter consists of a fixed set of tags (which grows and changes from time to time), the former allows almost arbitrary tags; the only restrictions are somewhat similar to restrictions concerning the names of variables for common programming languages. In addition to these tags XML has only a little more to offer. At first sight it might be hard for the experienced programmer to grasp the benefits of this ultimate form of freedom.

From the point of view of the EG-Models Server the – by far – most crucial point about XML is its built-in capability of getting rid of itself.

Access to Published Models

An import aspect of archived data is an easy and direct access to the data, and a simple citation and reference mechnism. Many of the drawbacks of classic print journals disappear in the realm of electronic versions. For example, it is possible to have a direct web link to an individual publication, or to allow a full-text keyword search.

Each model submission receives a unique identification number if the uploaded submission is complete and on first sight follows some basic requirements. Even rejected models which fail the later refereeing process will still keep their identification number for internal archiving purposes. Any published model on the server can uniquely be cited using this identification number.

The identification number has the form *xxxx.yy.zzz* where the four digits *xxxx* are the year of submission, the two digits yy the month of submission, and the three digits zzz are an internal counter of the received submissions starting with *001* each month and incrementing with each model. For example, the retract of Brakke has identification number *2001.01.042* which is the 42-nd model received in January 2001.

The identification number is also a key part of a unique URL which enables a direct internet access to a published model. Simply, attach the identification number to the EG-Model domain as in the following example:

http://www.eg-models.de/2001.01.042/

The EG-Models data base stores further keywords which allow providing various other ways to view and access the model collection. For example, a directory like browsing is based on a hierarchical set of subject keywords, or search robots may find models based on different search criteria.

Samples

This section presents some models published on the EG-Models server to give an overview of the variety of mathematical topics and to discuss various technical and computational aspects.

Deformation Retract Minimal Surface

This model [1] shown in Figure 3 by Ken Brakke analyses the tricky question of exactly what minimization problem a soap film solves. Reifenberg considered soap films to be sets that spanned a boundary in a homological sense. Adams described this soap film as an example of a set we would like to call a soap film, but which does not span its boundary in any homological sense. This film, on an unknotted wire, has a deformation retract to the wire. A retract from a set Y (the film) to a subset X (the wire) is a continuous map from Y to X which is the identity map on X. A deformation retract of Y to X is a homotopy from the identity map on Y to a retract of Y to X, with the image of the homotopy remaining in Y. It is, in fact, a strong deformation retract since the homotopy can be chosen to always be the identity map on the wire. The existence of this deformation retract shows that the soap film cannot be said to span the wire in any homological sense (at least if one stays with a wire-embedded-in-Euclidean-space model of soap films).

The data set of the model was produced by Brakke with inhouse software, and exported in the JVX format. Additionally, this model publication is accompanied with an interactive Java applet showing the actual homotopy as a user-controlla-

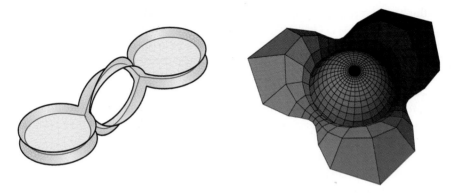

Fig. 3. Deformation retract minimal surface (left) and the Darboux transform of a discrete spherical isothermic net (right).

ble animation. Such additional interactivity may be supplied by an author by adding files in the category FILE_OTHER. These data belongs to the reviewed model submission, but the EG-Models managers provide no guarantee that these files are maintained. In the worst case, files in this category may become obsolete or may even stop working, for example, if a certain version of Java is no longer supported by major browsers. In contrast, the files in all other categories will be maintained by the managers of the EG-Models archive. For example if the specification of XML changes at some future time, then affected files of EG-Models publications may be updated into a different format.

Darboux Transform of a Discrete Spherical Isothermic Net

Smooth surfaces of constant mean curvature 1 in hyperbolic space can be characterized by the fact that a suitable Darboux transform (by means of the conformal Gauss map) yields the hyperbolic Gauss map. This provides one (of at least two) possibilities to define discrete horospherical nets – as analogs of smooth cmc-1 surfaces in hyperbolic space – as special discrete isothermic nets: note that the hyperbolic Gauss map, being part of the definition, determines the hyperbolic geometry the surface is horospherical in as a subgeometry of Moebius geometry. The displayed model [6] shown in Figure 3 by Jeromin was obtained as a Darboux transform of a spherical discrete isothermic net with high symmetry. It therefore is a horospherical net, and can be considered as a discrete analog of a smooth surface of constant mean curvature 1 in hyperbolic space: in the picture, the sphere at infinity of hyperbolic space sits inside the surface (the surface having two ends) – the standard Poincare ball model of hyperbolic space is obtained by inverting the configuration at the infinity sphere.

Simple Algebraic Singularities

This collection of algebraic surfaces [9] by Richard Morris presents the simplest types of singularities which can occur for functions from $\mathbb{R}^3$ to $\mathbb{R}$. The types consist of two infinite sequences and three special cases. The zero sets of the simplest of these singularities are presented here. All models have been produced by an applet based on the JavaView API [11] and a server based program which calculates the zero set of a given function. The server is adapted from the program in the LSMP package. The numerical algorithm has been constructed to try to get accurate representations of the singular points in the surfaces.

For the purpose of this collection the models have been cleaned up by hand to give topologically accurate representations of the singular points. These files have been hand edited to ensure that the boundaries are correct and that they are topologically correct around the singular point.

Densest Lattice Packing of a Truncated Dodecahedron

For a given convex body the lattice packing problem is the task to find a lattice of minimal determinant such that two different lattice translates of the body have no interior points in common. The ratio of the volume of the body to the determinant of such an optimal lattice is called the density of a densest lattice packing and it can be interpreted as the maximal proportion of the space that can be occupied by non-overlapping lattice translates of the body. The example in Figure 4 shows a truncated dodecahedron [5] from a collection of models by Martin Henk where the density of a densest lattice packing was calculated with the algorithm of Betke and Henk. The density is equal to $(37 + 5\sqrt{5})/(24\sqrt{5})$, and the 12 points in the picture show the lattice points of a critical lattice lying in the boundary.

Counterexample to the Maximum Principle of Discrete Minimal Surfaces

The authors define discrete minimal surfaces as piecewise linear continuous triangulated surfaces that are critical for the area functional with respect to all variations through surfaces of the same type that preserve the simplicial structure. Unlike smooth minimal surfaces, a discrete minimal surface might not lie in the convex hull of its boundary, see Figure 4. This example [12] by Konrad Polthier and Wayne Rossman is a simple case of such a surface, and thereby disproves existence of a discrete version of the convex hull principle for discrete minimal surfaces.

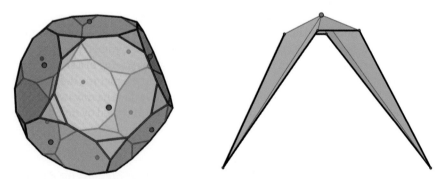

Fig. 4. Densest lattice packing of a truncated dodecahedron (left) and a counterexample to the maximum principle of discrete minimal surfaces (right).

More general, the maximum principle for solutions of elliptic partial differential equations says, that if two solutions are tangent at a common point and one solution lies on one side of the other solution then both solutions are identical. This maximum principle does not hold for discrete minimal surfaces as this model demonstrates: the central vertex lies outside the convex hull of its boundary, therefore, any xy-planar triangulation, which is certainly discrete minimal, lies one side of the example. For example, consider the planar triangulation obtained from projecting the surface onto the xy-plane.

Sharir's Cube

A three-dimensional convex polytope is a *cube* if its face lattice is isomorphic to the standard cube, which arises as the convex hull of all three-dimensional vectors with 0/1-coordinates. Many examples of strange cubes are known. Some of them turn out to be surprisingly di¢cult to handle for certain (Simplex type) algorithms for linear programming. Each quadrangular facet F^+ of a cube has an *opposite* facet F^-, which is characterized by the condition that F^+ and F^- do not share any vertex. In the standard cube any two opposite facets are parallel, but it is easy to deform a cube such this is no longer true. Micha Sharir asked the question whether it is possible to have a cube such that each facet is *perpendicular* to its opposite. Ziegler gave an explicit construction [15] which establishes a positive answer. The model is shown in Figure 5.

For this kind of model it is essential to have exact data. Therefore, the master file is in polymake [3] format, which allows arbitrary precision rational coordinates. Any representation of a polytope with floating point coordinates which fail to satisfy the numerical conditions by a small margin would be useless.

A Secondary Polytope

An ingenious construction of Gelfand, Kapranov, and Zelevinsky [4] associates to each triangulation of a set of n points in d-space one point in n-space. The i-th coordinate of this new point is the total volume of all the simplices incident to the i-th point in the triangulation. The convex hull of these new points is called the *secondary polytope* of the point configuration. It has dimension $n - d - 1$, and its vertices correspond to the so-called regular triangulations of the point set.

Fig. 5. Explosion of the boundary of Sharir's Cube (left) and a secondary polytope with interior points (right).

Since it is a difficult task to enumerate all the triangulations of a given point configuration, it is computationally hard to produce non-trivial examples of secondary polytopes. Pfeifle used TOPCOM [13] and polymake to provide some examples, see [10] and Figure 5, which can now be used as a starting point for further investigations.

References

[1] K. Brakke. *Deformation retract soap film.* Electronic Geometry Models, 2001. http://www.eg-models.de/2001.01.042/.

[2] G. Fischer. *Mathematische Modelle/Mathematical Models.* Vieweg, Braunschweig, 1986.

[3] E. Gawrilow and M. Joswig. *polymake, version 1.4: a software package for analyzing convex polytopes,* 1997–2001. http://www.math.tu-berlin.de/diskregeom/polymake.

[4] I. Gelfand, M. Kapranov and A. Zelevinsky. *Discriminants, resultants, and multidimensional determinants.* Mathematics: Theory & Applications. Birkhäuser Verlag, 1994.

[5] M. Henk. *Densest lattice packing of a truncated dodecahedron. Electronic Geometry Models,* 2001. http://www.eg-models.de/2001.02.064/.

[6] U. Jeromin. *Darboux transform of a discrete spherical isothermic net.* Electronic Geometry Models, 2000. http://www.eg-models.de/2000.09.038/.

[7] M. Joswig and K. Polthier. *Digital models and computer assisted proofs.* EMS Newsletter, December, 2000.

[8] M. Joswig and K. Polthier. *Digitale geometrische Modelle.* DMV Mitteilungen, 4:20 – 22, 2000.

[9] R. Morris. *Singularities of algebraic functions in R 3.* Electronic Geometry Models, 2001. http://www.eg-models.de/2001.06.002/,...,2001.06.020/.

[10] J. Pfeifle. *Secondary polytope of a cyclic 8-polytope with 12 vertices.* Electronic Geometry Models, 2000. http://www.eg-models.de/2000.09.032/.

[11] K. Polthier, S. Khadem-Al-Charieh, E. Preuß, and U. Reitebuch. *JavaView Homepage,* 2001. http://www.javaview.de/.

[12] K. Polthier and W. Rossman. *Counterexample to the maximum principle for discrete minimal surfaces.* Electronic Geometry Models, 2001. http://www.eg-models.de/2000.11.040/.

[13] J. Rambau. *Topcom 0.9.0.* http://www.zib.de/rambau/TOPCOM/.

[14] M. Schilling. *Catalog Mathematischer Modelle Für Den Höheren Mathematischen Unterricht.* Leipzig, 1911.

[15] G. M. Ziegler. *Sharir's cube.* Electronic Geometry Models, 2000. http://www.eg-models.de/2000.09.028/.

Trees, Roots and a Brain:
A Metaphorical Foundation for Mathematical Art

John Sims

Introduction

The ideas and language of mathematics have played an important role in the development of the visual arts. Whether it is perspective geometry giving rise to academic realism, or the revolutionary work of Kandinsky, mathematical thinking and language have framed the art process by expanding the way we view and conceptualize reality. Furthermore, the aesthetics of mathematics has spirited art movements such as Constructivism as well as informed the minimalism and conceptual work of such artists as Max Bill, Sol LeWitt, Mel Bochner and Agnes Denes, who have questioned the very foundational relationship between art and idea. Delivering rich visualizations with a masterly aesthetic, contemporary artists such as Brent Collins, Helaman Ferguson and Tony Robbin continue the mathematics-art saga with forceful leadership.

As an emerging Math-Artist, I aim to advance a mathematical art that explores subject matters in a philosophical-metaphorical context. By cross-pollinating symbols and ideas from a multitude of spaces both cultural and mathematical, I seek to create works that promote a deeper visual and conceptual critique of objects of both nature and mind. I begin this project with a distillation model of vision that uses as parameters: nature, art and mathematics. Using a tree-root metaphor I will present three metaphorical-visual relationships between art and mathematics. These visual metaphors will become the foundational visual philosophy for a metaphor-based mathematical art.

Vision

Visual communication has been central in shaping human knowledge and experience. Whether it is a mural of Diego Rivera pictorializing factory workers or the graph of the sine function depicting the shape of periodicity, visual demonstrations and metaphors are central to human consciousness and capacity to communicate ideas and emotions. And being that mathematics and art are essential parameters of human consciousness as well as an important element of visual communication, I am interested in various ways pictorial information can convey different kinds of knowledge. Consider the following distillation of vision: retinal, re-creative, and mathematical. The retinal vision is the near passive reception of images of nature as perceived by the human nervous system. The re-creative is the artist's vision and interpretation of an object. The mathematical will be the vision of geometry and structure that laced real and abstract objects. This leads to a trichonomy that parallels the critical points in Western Art: realism-expressionism-conceptual art.

| Retinal | Re-creative | Mathematical |

Tree

We begin with using the concept of a tree as an example. The above images show the retinal, re-creative and the mathematical representations of tree. As we travel from the retinal to the mathematical, we go from the concrete to the abstract and from realism to conceptual. These critical elements will become symbols in a co-ordinate system, where mathematics, art and nature are represented on the x, y and z-axes. I will use these trees to construct visual metaphors that show the relationship between mathematics and art.

Mathematics and Art: A Tree-Root Metaphor

Our discussion begins with the concept of trees viewed as objects of nature, mathematics and art. Combining the mathematization of a tree, a Pythagorean fractal tree with the drawing of a tree, we create *Square Roots of Tree* and *Tree Root of a Fractal*. These images, using a tree-root metaphor, illustrate a philosophical connection between mathematics and art, leading us to examine the following questions: What is the structural connection between Mathematics and Art?

There are various ways to view mathematics. Overall, the essence of mathematics manifests itself in the capacity to see and show structural relationships. Through its symbols and visual grammar, mathematics as language reflects with an aesthetic of minimalism a strong desire for clear communication. Viewed as conceptual technology, mathematics, being the foundation for modern science, is also an indispensable tool for the other divisions of human knowledge. As conceptual art, mathematics and its process embrace the beauty of abstract ideas and the celebration of structure.

While the visual arts have become a powerful agency of expression, its connection to visual mathematics is undeniable. The geometry of the times frames the conceptual and structural landscape of expression as symbolic in the Square Roots of a Tree, which presents mathematics as foundational and the conceptual root of form. While the art process harnesses technology to visually communicate ideas, it conjoins with the mathematical process to produce visual mathematics. To see mathematically one draws from creativity and intuition, as in the case with the art process itself. *Tree Root of a Fractal,* also shows how the latent geometry of nature can inspire and support abstraction. These images essentially show an interdependency between mathematics and art.

Square Roots of a Tree

The Mathematical Art Brain

Rotating the *Square Roots of Tree* 90 degrees, we get a brain, with the geometry of the Pythagorean tree on the left side and the drawing on the right side. This image is a metaphor for a conceptual

Tree Root of a Fractal

165

nervous system that facilitates experiencing the world through mathematics-art aesthetics. This image represents the foundation in which I as a math-artist seek to

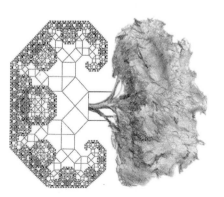

Mathematical Art Brain

create works of visual metaphorical based art. As with M.C. Escher, I wish to create narratives that blend symbols and visual language of everyday life with mathematical visual references for the purpose of promoting a discourse about the world around us. More generally, this conceptual brain is a metaphor for a more holistic way of processing knowledge and experience. This model should challenge us to re-think our ideas about mathematics and art, art criticism, and art and mathematics education.

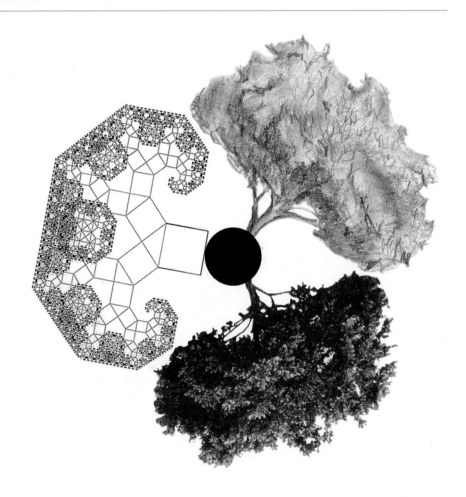

Constructing the 3-D Tree

What does a tree "look" like through the vision of a conceptual nervous system guided by the *Mathematical Art Brain*? The *3-D Tree* is a tree recorded through the experience of such a brain. We discuss this tree as a conceptual distillation of three essential components of consciousness: abstraction, expression, and perception, extending past cubism into a proto-conceptual animation of a tree. By adding the retinal part to the previous visual metaphors, we get a triad that acts as a conceptual coordinate system that re-orients the notion of a tree. The tree-root relationship is still present, with the retinal and expressive being collateral roots for the mathematical, the expressive and mathematical being roots for the retinal, and mathematical and retinal being the roots for the expressive. This image offers rich possibilities into the metaphorical projections of meanings regarding the relationship between mathematics, art and nature. It embodies a new way to think about trees building on the traditions of perspective and cubism.

Trans-Cubism

Cubism is concerned with the spatial com-pactification of an object leading to a geo-metric narrative that induces a space-time continuum. Breaking from the projective ge-ometry of Renaissance paintings, cubism seeks to introduce more options allowing one to visualize the fascinating tension be-tween sub-structure and global structure and to capture the sequentiality of a proto-animation narrative. By thinking of objects in relationship to their mathematical, re-cre-ative and physical dimensions, we can create a visual distillation that extends beyond the geometrical cutting of traditional cubism and investigate some of the metaphorical in-formation that can be used to understand the essence of the objects on a deeper level. The trans-cubism ideas behind *3-D Tree* can be applied to other objects as well. It is interest-ing to note that not all objects submit to a conceptual distillation as a tree might. For example, things such as emotions or a fork resist easy mathematical descriptions and therefore would not offer an elegant distilla-tion.

Cellular Forest

Since trees usually live with other trees in a forest whose arrangement is an organic pat-tern directed by the whim of nature, it would be desirous to tile a region of space with the *3-D Tree* of different sizes and rotations. The placements of these trees are done without a conscious plan to obtain order but to produce an intuitive sense of balance. Once organized into a cellular form, the trees are connected together with lines with an emphasis on shortest distances. This gives us the *Cellular Forest*, which now can be used to tile a bigger region according to the geometry of its own internal structure. This begins a fractal pro-cess that can be repeated.

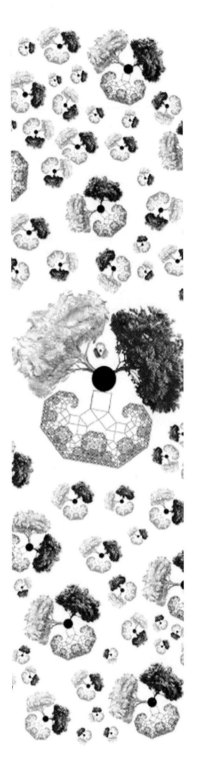

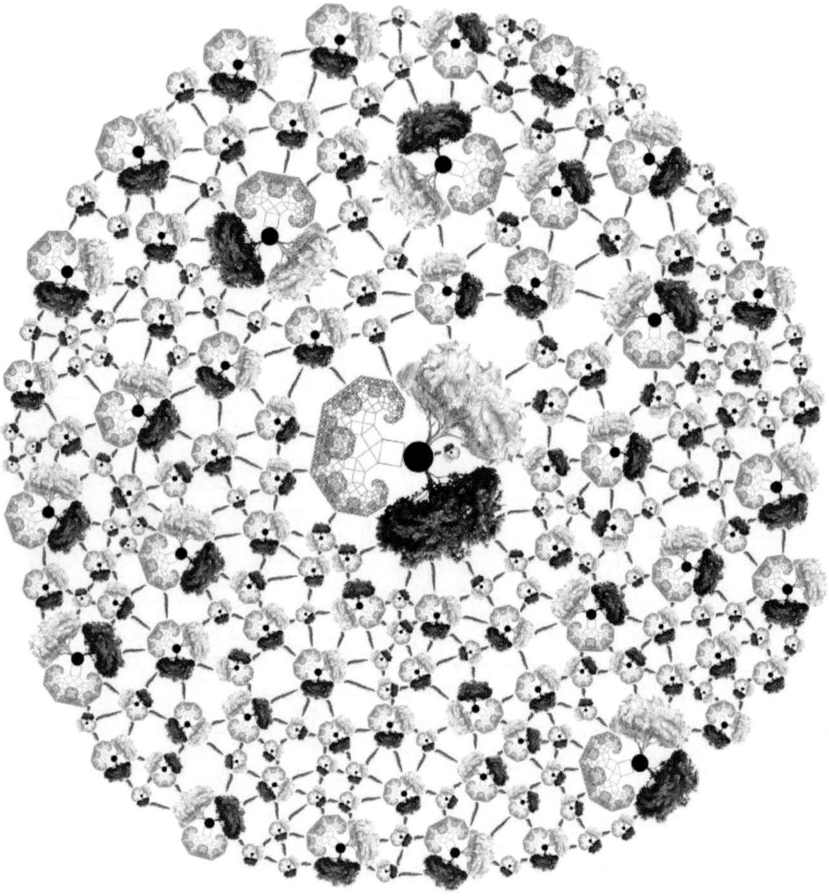

Conclusion

Mathematics and art are important dimensions of human cognition connected by shared and complimentary modes of processing. Mathematical Art sitting in the intersection of these dimensions embodies the ever so powerful tension between creativity and analysis, concrete and the abstract, and the unique and the universal. I have attempted to construct a theoretical framework with which to create mathematical-art works that go beyond the traditional ways mathematics has been encoded into the art process. By speaking to the human experience as it relates to perception, expression and mathematics, and with the aid of a tree-root metaphor I have tried to illustrate an interdependent relationship between mathematics, art and nature. The image *Square Roots of a Tree* speaks to the mathematics as a support structure for art. With *Mathematical Art Brain,* we have a symbol for a metaphysical nervous system that represents an evolution in human vision combining elements of realism, expression and conceptualism in a trans-cubism way. And *Tree Roots of a Fractal,* shows us how nature can inspire and create mathematics. The following summarizes these relationships in the triptych entitled *Mathematical Art Philosophy 101.*

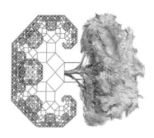
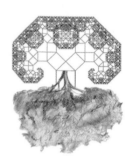

Square Roots of a Tree　　　Mathematical Art Brain　　　Tree Root of a Fractal

The visual metaphors I have presented here are the beginning of a body of work, started with M.C. Escher, that combines the spirits of mathematics, art and nature in a variety of ways. Although recent developments in western culture have put mathematics and art at odds with each other, development in technology and attention towards visual mathematics have created incredible opportunity to advance the genre of Mathematical Art. With this in mind, it is also important that an art criticism be developed that is sensitive to emerging mathematics-art aesthetics. It is important as well that artists be encouraged to embrace the essence of mathematics and recognize it as a great conceptual tool to develop new knowledge about the world we occupy. As Mathematical Art matures, particularly in a narrative context, it will become a powerful metaphysical source of insight into the beautiful complexities of the nature and the mind.

References

[1]   Berger, John, *Ways of Seeing*, BCC, London, UK and Penguin Books, NY, 1972.

[2]   Emmer, Michele, *The Visual Mind: Art and Mathematics*, MIT Press, 1993.

[3]   Levin, David Michael, Editor of Sites of Vision: *The Discursive Construction of Sight in the history of Philosophy*, MIT Press, 1997.

[4]   Taylor, Harold, *Art and the Intellect*, The Museum of Modern of Art, New York, 1960.

[5]   Whitehead, Alfred North, *Symbolism: its meaning and effect*, Capricorn Books, NY, 1927

Mathematics, Literature and Cinema

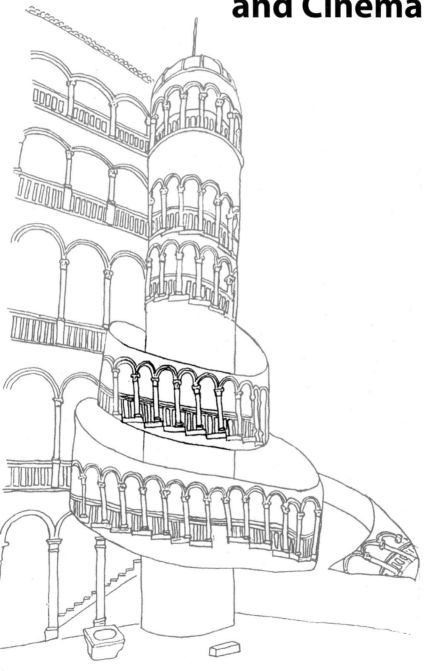

Mathematics, Literature and Cinema

Michele Emmer

In recent years, the public have surely associated the image of a mathematician with that of Russell Crowe in the film, *A Beautiful Mind*. [1] Whereas until a few years ago, if one asked someone to name a mathematician, most were likely to say Archimedes or Pythagoras, now the name John Nash is very well known. Obviously it would be an entirely different matter were one to ask what kind of mathematics Nash did, or what the De Giorgi-Nash theorem is about. Certainly, films are not made for a population of mathematicians. Until a few years ago, mathematicians were practically non-existent at the cinema, and if they did appear, it was as a 'baddy'.

After a great success in the theatre, the film adaptation of *Proof* by the comedian David Auburn arrived on the screen in 2004. Anthony Hopkins plays the lead role, John Madden directs and the screenplay is by Rebecca Miller. What Anthony Hopkins said in an interview was strange: "To tell the truth, I did not get on well at school, I didn't have a proper education, I never went to university. And I couldn't for the life of me ever have been a teacher, I'm too stupid." However, evidently he has the physique and the look of the role, of a genius mathematician, as is required by the protagonist in *Proof*.

Yet another film, "21 gramms" [2], with Sean Penn playing the role of a mathematician, won the 2003 Volpi Cup for best male actor at the Venice film festival and was nominated for an Oscar. A proper numerical film made in different installments by Peter Greenaway is *The Tulse Luper Suitcase* [3]. Greenaway has always been fascinated by numbers and has often used actual number grids in the making of his films. Sometimes the films are made using the number 100, as in the famous *Drowning By Numbers* (which was released as *Water Games* in Italy as the word *number* was not yet considered appealing!), in which the first one hundred whole numbers were hidden in every scene. The film project, the first part of which was shown in Italian cinemas in 2004, is composed of three episodes, and is based on the atomic number of uranium, 92. That's how many characters and how many objects are used to describe the world, objects that appear every now and then throughout the film.

Greenaway has written a long article in which he describes how he first came to be interested in numbers and number grids for use in his films. The article is called *How to Make A Film* [4] and Greenaway is unambiguous. It wasn't by chance that he called his book *Fear of Drowning by Numbers* (*Fear of Numbers* in Italian), which had the subtitle *100 Thoughts On Cinema*. What is the special role of numbers in cinema? "Counting is the simplest and most primitive form of narration – 1 2 3 4 5 6 7 8 9 10 – a story with a start, a middle and an end, and a sense of progression – which culminates in a finale of two digits – a purpose fulfilled, a conclusion arrived at:"

In these last few years many books that talk of mathematics and mathematicians have had great success around the world. Among others, it will suffice to mention that of Simon Singh, dedicated to the proof of Fermat's Last Theorem.

In his book *Mathematics in Western Culture* [6], Morris Kline wrote "It is even less widely known that mathematics has determined the direction and content of philosophical thought, has destroyed and rebuilt religious doctrine, has supplied substance to economics and political theories, has fashioned major painting, musical, architectural, and literary styles, has fathered our logic, and has furnished the best answers we have to fundamental questions about the nature of man and his universe ... Finally, as an incomparably fine human achievement, mathematics offers satisfactions and aesthetic values at least equal to those offered by any other branches of our culture."

Many things have changed in the years since Kline wrote this, and a more widespread knowledge of mathematics can only be good for culture, be it scientific, literary or artistic.

References

[1]   Emmer, M.: *A Beautiful Mind,* recensione, B.U.M.I., series VIII, vol. IV-A, August 2001, p. 331–339.
[2]   Inarritu, A. G., director, *21 Gramms,* actors: Sean Penn, Benicio del Toro, story and script by A. G. Inarritu (2003)
[3]   Greenaway, P., director, *The Tulse Luper Suitcase,* actors: J.J.Field, Drew Mulligan, Valentina Cervi, Isabella Rossellini, Tom Bower, story and script by Peter Greenaway.
[4]   Greenaway, P., *Come costruire un film,* in *Matematica e cultura 2000,* ed. M. Emmer, Springer, Milano, 2000, p. 159–171; English edition, *Mathematics and Culture 1,* Springer, Berlin (2004)
[5]   Greenaway, P., *Paura dei numeri,* Editrice Il Castoro, (1996)
[6]   M. Kline, *Mathematics in Western Culture,* Oxford University Press, New York, 1953
[7]   Kuhn, H.: *La matematica al cinema arte: analisi di un caso esemplare,* in *Matematica e cultura 2003,* ed. M. Emmer, Springer, Milano, 2003, p. 135–148.

Euclid's Poetics: An examination of the similarity between narrative and proof

APOSTOLOS DOXIADIS

Introduction

I want to state and briefly explore what I believe to be strong structural analogies between making narratives and proving mathematical theorems – analogies a mathematician might be tempted to called 'isomorphisms', i.e. one-to-one correspondences of the elements of two sets that, additionally, preserve their structure. My thesis does not lay claim to the rigor of a purely mathematical result. But I hope that by being even approximately accurate, it points in an interesting direction.

In graphic notation, I want to show that:

The idea of this isomorphism, which I'll call F, has been on my mind for quite a while, but really began to solidify (the case of a drop making the glass overflow, really) when I heard a reader of my novel *Uncle Petros and Goldbach's Conjecture*, (1) comment that the story it tells "unfolds much as solving a mathematical problem."

Pursuing this analogy, I would like to give arguments for my main thesis much as 'solving a mathematical problem'. The technique I will use will be an application of the transitive quality: to prove that A is equal (or isomorphic) to C, it is enough to prove that both are independently equal (or isomorphic) to a certain B.

Thus, A=B & B=C implies that A=C

This common reference point is in the case of my thesis a *spatial analogy*, which I believe underlies both narrative and proof.

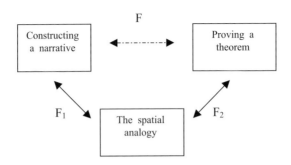

Thus, I will try and show the structural equivalence, F, between constructing a narrative and proving a theorem by showing how both of these are independently equivalent to a spatial model, the equivalences F1 and F2. The transitive quality then guarantees that F = F1 & F2

2. The underlying spatial metaphor of narrative

Since Aristotle's *Poetics* there has been an attempt to find universal laws underlying the structure of narrative. Interestingly, the most important insights were achieved in the twentieth century by theorists operating outside the field of literary studies proper. The Russian folklorist Vladimir Propp, in a seminal essay (2) finds that the so-called 'magical folktale' always conforms to a particular structure involving standard 'functions' (his term) that can range over a set of variables, giving different versions of a more or less constant underlying structure. Roughly, this structure is:

- The hero lives in a condition of stability.
- Something upsets this condition.
- The hero embarks on a journey to restore stability.
- He faces challenges assisted by a 'magical assistant', who is often an animal.
- The final challenge(-s) are successfully faced.
- The hero comes to a higher state of stability, because of his actions.

What is important to my thesis is that underlying all these phases there is a *journey* to (geographical) points of which every phase of the journey can be associated: crucial encounters, acquisition of information or objects, challenges, fights, magical events, revelations, etc., all can be laid out, as it were, on a map, every step of the hero having a spatial analogue. These are often charged (but don't need to be) with a metaphorical resonance. Thus, advancement of the story is forward movement, decisions are cross-roads, the narrative goal is also a physical destination, etc. and there is of course the full process of coming full circle, from stability, to instability, to stability.

Anthropologists and historians of religion later generalized this kind of narrative structure, speaking of the 'quest of the hero' as the archetypal myth, a thesis presented by Joseph Campbell's famous book, *The hero with a thousand faces*. In more recent years, help has also come from the unlikeliest place: Hollywood. Trying to codify the underlying structure of a film-script, scriptwriting teachers and 'script-doctors' (sic!) have resorted to Propp and Campbell, seeing in the pattern of the quest myth almost universal validity, as the sort of *Ur*-story, the primal, archetypal narrative. And although their insights have resulted, largely, in an endless torrent of highly similar and very often vacuous films, their analysis has a lot going for it. By looking at countless stories, whether they be recorded on film, the page, or retold by the human voice, one can see that most of them conform essentially to this pattern: **a hero wants something and embarks on an adventure-laden journey to get it.** This 'something' that the hero wants (be it a person, an idea, a material object, whatever) is the goal of the journey or, speaking spatially, its destination. If we further generalize the definition of the quest myth and replace the '**hero wants something**' with the '**hero wants something** *or* **the author wants something for him/her**', then this encompasses practically all narratives or, to be exact, practically all *simple* or *elementary* narratives, as often a longer narrative, say a novel by Dickens, is made up of a combination of many simpler ones.

Let us look at some famous examples of heroic goals/destinations:

HERO	GOAL
Ulysses	Ithaca
Oedipus	Cure of the plague
Lancelot	Guinevere, the Grail
Hamlet	To revenge father
Romeo	Juliet
Juliet	Romeo
Jay Gatsby	Daisy
The three sisters (Chekhov)	Moscow
The old man (Hemingway)	The fish

Now, the hero's journey may be very literal (as, say, in the *Odyssey*) or very metaphorical (as in T. S. Eliot's *Four Quartets*) and is often both at the same time, as for example in the medieval legend of the Grail. But whether metaphorical or literal or both, what's essential to our discussion is, again, that the hero's journey can be mapped (interestingly, a geographical as well as mathematical expression) i.e. can be given precise spatial form, even if this 'space' can also be immaterial, as is, say, the world of memory or imagination.

As to the hero reaching the destination, literature has gone a long way beyond the alternatives of the traditional quest myth, a Gilgamesh, Odysseus or Parcifal, and their various versions of a 'happy end'. The reaching of the goal (destination) can take many different forms, as for example:

- The goal is reached and this fulfills the hero's need.
- The goal is reached but the hero finds he is disappointed with it.
- The goal is reached but then the hero realizes a new goal lies ahead and thus embarks on a new journey.
- The goal is reached but this only makes the hero realize the importance of the journey over the goal.
- The goal is only partially reached and the hero realizes and accepts this.
- The goal is only partially reached, the hero realizes and does not accept this.
- The goal is not reached, and this makes the hero sad.
- The goal is not reached but that's alright, because the hero has reached a new insight.

And so on. To summarize: **almost all stories have to do with a hero wanting to (or the author wanting the hero to) get something**. This can almost always be translated, structurally, to the wanting to get *somewhere*, by following a certain course, literal or metaphorical. Thus, any narrative can be represented as a journey, with a beginning (B) and an end (E) with various forces (arrows) operating as either 'helpers' (Propp's term) external or internal, or obstacles, influencing the course of the hero's progress. Dotted lines here indicate 'the roads not taken', in T. S. Eliot's famous phrase, alternative courses the hero did not finally choose.

This more or less settles the first part of our argument, i.e. that there exists an isomorphism, which we called F_1, between narrative and a spatial model.

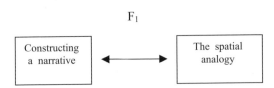

3. The underlying spatial metaphor of mathematical proof

I first hit upon the idea of the spatial analogy also underlying mathematical proof when reading in the *Homilies on the Hexaemeron* of the fourth century Christian theologian Saint Basil, his wonderful insight that the dog (yes, the *dog*) can be credited with the invention of the mathematical method of *reductio ad absurdum*.

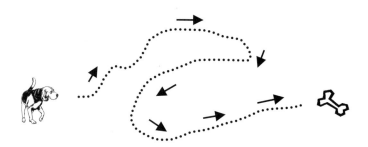

You see, when a dog searches for the desired object (bone) he will begin to sniff a likely trail and if disappointed will retrace his steps somewhat and start off in a new direction. Obviously, this brought to Saint Basil's mind the method of someone like Euclid, when saying: 'let us assume that the primes are finite, and see what happens'. (As is well known, Euclid then follows the consequences of this hypothesis and since it brings him to a contradiction applies the principle of the excluded middle to conclude, in more modern terminology, that if 'not-P is false, then P must be true'; or, in our example, that since the primes cannot be finite, they must be infinite.) But this too can be expressed with a simple algorithm, which is really spatial: 'at a crossroads, forking into roads A and B of which one leads to a cul-de-sac and the other to the treasure, first take A. If it leads to a cul-de-sac, then retrace his steps, take B and be led to the treasure with certainty.'

Here, we must make the crucial distinction between the proof of a mathematical theorem as it is experienced by a student/reader studying an already discovered, published result, and as it is was originally established by the mathematician(-s) who discovered it. It is this second viewpoint that is more interesting, although of course a published proof may contain, in an often indirect sense, part of the intellectual adventure of its completion. The process of proof can be very simple (again, see Euclid's proof of the infinity of primes) but it can also be long, arduous, complicated and multi-faceted. A good example of this is Andrew Wiles famous proof of Fermat's Last Theorem, which was the culmination of a very long process lasting a few decades (or centuries, if you want to go back to Galois and the origins of modern algebra) and was successively created (although with no clear end in sight, for a long while) by a number of mathematicians, among them Taniyama, Shimura, Weil, Frey, Ribet, and a few more with Wiles providing the final integrative thrust that brought the various threads together. (3)

Like a narrative, such a process of gradual discovery, whether long or short, complex or simple, can be mapped, i.e. it can be given a spatial form. In fact, more or less everything we said comparing the narrative to the spatial model holds also true of the process of mathematical proof.

Let us investigate this point: a mathematician starts out wanting to prove a proposition, which is really the end of his *destination* (Of course, he may also start out, like a hero in some modernist fiction, merely by fooling around with ideas, with no destination, just a general sense of ennui leading to curiosity, leading to questions.). Here are some examples:

THE HERO	THE GOAL
Euclid	The primes are infinite
Newton/Leibniz	How to find gradient of curves
Evariste Galois	The solution of 5th degree equations
Henri Poincaré	The Three Body Problem
Atle Selberg	Elementary proof of the Prime Number Theorem
Stephen Smale	The higher-dimensional Poincaré Conjecture
Andrew Wiles	$x^n + y^n = z^n$ admits no integer solutions for $n>2$

Most aspects of the process of proof will admit a spatial correlative:

- The mathematician moves forward (often backwards, or sideways) in logical space, searching this way and that.
- The mathematician may take advantage of road maps, of greater (already proven results) or lesser (conjectures) accuracy.
- The mathematician will face challenges, disappointments, will win some fights (intermediate results) and lose some (cul-de-sacs), may often change direction, will be assisted by 'magical assistants' (mentors, colleagues, the accumulated knowledge of the past), may employ powerful talismans or weapons (new methods) and will finally (in a 'happy end' scenario) reach his destination – i.e. prove the desired theorem. All these have their analogues in logical space, which we can envision as a decision-studded magical, metaphorical forest.

Of course, the happy ending is not obligatory. The mathematician may not reach his goal, or find it not at all similar to his expectations (Nagata working on Hilbert's Fourteenth Problem only to finally prove it false), or, again like some modernist hero, may think that he has arrived, while he really hasn't – like Fermat thinking he had proved his theorem when (we think now) he hadn't.

In fact, the possible outcomes of his spatial progress into the forest (maze, labyrinth, whatever) may end in some of the various ways that we thought were reserved for fiction. Thus, for example – and I am here taking similar options to those presented earlier (for 'mathematician', read 'hero'):

- The goal reached and this fulfills the mathematician's need (e.g. Euclid and the infinity of primes.)
- The goal is reached but the mathematician and/or others are is disappointed with it (e.g., or the proof of the famous Four Color Theorem, which was so cumbersome that some do not accept is a proof.)
- The goal is reached but then the mathematician realizes a new goal lies ahead and thus embarks on a new journey (the proof of a theorem points to a much more important result).
- The goal is not reached but this only makes the mathematician realize the importance of the journey over the goal (while trying to study the distribution of primes, Riemann invents his zeta function.)
- The goal is only partially reached and the mathematican realizes and accepts this (proofs that don't manage the full result but a weaker version of it, e.g. Jing-Run Chen's proof that every even number is the sum of a prime and an almost prime – a weaker version of Goldbach's Conjecture).
- The goal is not reached but that's alright, because the mathematician has reached a new insight (Galois failing to find a formula to solve the quintic equation, but discovering group theory and a lot more on the way) .
- The goal is only partially reached, the mathematician realizes and does not accept this. (Alas, countless examples.)
- The goal is not reached, and this makes the hero sad. (The same.)

These arguments seem to take care of the second part of our proof (oh, call it 'argument', if 'proof' sounds to strong), demonstrating the isomorphism:

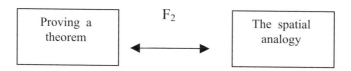

4. Conclusion

We now seem to have come to the desired point, completing our with the isomorphism F, between narratives and proofs, by virtue of the transitive quality (F_1&F_2 implies F):

I suspect that this analogy (isomorphism) does not seem too interesting to a mathematician or, not to be unfair to the more poetically inclined, it doesn't seem *useful*. Knowing that proving a theorem may look somewhat like the unfolding of a story is certainly no help to a mathematician in proving new theorems – and, like it all not, this is the prime criterion of usefulness to mathematicians. But the analogy may be more useful to people dealing with narratives. Although it will not answer Hollywood's dream of a magical formula to create more interesting narratives, it does point at a handy formalism, and at analogies which can provoke a storyteller's thoughts.

And what more can a storyteller want?

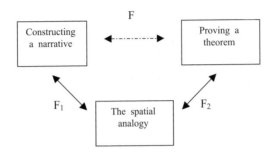

References:

[1] A. Doxiadis *Uncle petros and Goldbach's Conjecture,* Faber & Faber, London (2000).
[2] Propp, Vladimir, *The Morphology of the Folktale,* (1968) translated by Laurence Scott, University of Texas.
[3] Simon Singh *Fermat's Enigma,* Anchor/Doubleday, New York, 1997.

The Rise of Narrative Non-Fiction

Simon Singh

Traditionally popular science writers have put the emphasis on explanation, concentrating on conveying to the reader an understanding of scientific concepts. There have been numerous successful books that follow this archetype, including the recently published *The Elegant Universe* by Brian Greene. The book has been successful because it clearly explains the ideas of relativity and quantum physics and how string theory offers hope of unifying these two models of the universe. The public will always appreciate any book that successfully explains the latest scientific understanding of the universe.

However, the last five years have witnessed the burgeoning of a new type of science writing, so-called narrative non-fiction, in which the emphasis is not solely on the on the explanation of science. Instead, the author also writes about the scientists, their motives, adversities and triumphs. All of this is framed within an overarching narrative. These books still explain the science, but they also tell the tale of a scientific discovery or have a biographical thread.

The ratio of explanation to story in science writing has a spectrum that ranges from academic papers (dominated by explanation) to text books to traditional science writing to narrative non-fiction (even balance between explanation and story). It is even possible to go far beyond narrative non-fiction, where we find fiction based on scientific or mathematical themes. In these books the story is naturally more important than any explanation of scientific concepts, but they do explain what drives scientists, describing the culture and atmosphere of scientific research. Recently there have been several fictional books about mathematics namely *Uncle Petros* and *Goldbach's Conjecture* by Apostolos Doxiadis and *The Parrot's Theorem* by Denis Guedj.

Arguably the trend towards narrative non-fiction began with Dava Sobel's *Longitude,* a description of the invention of the marine chronometer, which also tells the story of its inventor John Harrison, who had to battle with the establishment in order to get his breakthrough recognised and adopted. Subsequently, many other books have been categorised as narrative non-fiction, including my own books, *Fermat's Last Theorem* and *The Code Book*.

The Code Book is a history of cryptography. We can see the difference between traditional non-fiction and narrative non-fiction by examining chapter 6, in which I discuss a system of encryption called public key cryptography, one of the greatest cryptographic developments in history. Traditional non-fiction would concentrate on explaining the mathematics and mechanics of public key cryptography. It is a fantastic, counter-intuitive and brilliant concept, so naturally readers would appreciate a clear explanation. In *The Code Book* I do, of course, explain the concept of public key cryptography, a system which is powerful because it allows two people (known as Alice and Bob) to communicate securely with each other without having previously agreed or exchanged a key (the recipe for encrypting and decrypting). The following excerpt gives an analogy for public key cryptography:

Start Quote

This anecdote concerns a country where the postal system is completely immoral, because postal employees will read any unprotected correspondence. One day, Alice wants to send an intensely personal message to Bob, and so she puts it inside an iron box, closes it and secures it with a padlock and key. She puts the padlocked box in the post and keeps the key. However, when the box reaches Bob, he is unable to open it, because he does not have the key. Alice might consider putting the key inside another box, padlocking it, and sending it to Bob, but without the key to the second padlock he is unable to open the second box, and so he cannot obtain the key that opens the first box. The only way around the problem seems to be for Alice to make a copy of her key and give it to Bob in advance when they meet for coffee. We are back to the same old problem of key distribution. Avoiding key distribution seems logically impossible – surely, if Alice wants to lock something in a box so that only Bob can open it, then she must give him a copy of the key. Or, in terms of cryptography, if Alice wants to encipher a message so that only Bob can decipher it, then she must give him a copy of the key. Key exchange is an inevitable part of encipherment … or is it?

Picture the following scenario. As before, Alice wants to send an intensely personal message to Bob. Again, she puts her secret message in the box, padlocks it and sends it to Bob. When the box arrives, Bob adds his own padlock and sends the box back to Alice. When Alice receives the box, it is now secured by two padlocks. She removes her own padlock, leaving just Bob's padlock to secure the box. Finally, she sends the box back to Bob, who can now open the box, because the it is only secured with his own padlock, and he has the key to his own padlock.

End Quote

By performing a triple exchange with two padlocks it seems as though key distribution is not an inevitable component of encryption. The book goes on to explain the evolution of this concept and the eventual mathematical implementation. Furthermore, *The Code Book,* goes on to tell the intriguing story that surrounds the invention of public key cryptography, which is why it has been labeled an example of narrative non-fiction.

For example, *The Code Book* describes the political, social and technological circumstances that motivated the development of public key cryptography. It then introduces the three scientists who made the crucial breakthrough, namely Whitfield Diffie, Martin Hellman and Ralph Merkle. The book describes their backgrounds, their struggles, and the moment of their breakthrough. For example, one section describes Hellman's childhood as a Jewish kid growing up in a Catholic neighborhood of New York, which contributed to his independent attitude. Having been frustrated at not being like the other kids (e.g. not celebrating Christmas), he decided that it was better to be different, and radical thinking was one facet of being different.

Diffie, Hellman and Merkle developed the concept of public key cryptography, but they were unable to construct the mathematics required to make it work in practice. The *Code Book* tells the story of another trio (Rivest, Shamir and Adelman, or RSA) who were able to complete the development of public key cryptography. The book describes how the RSA cipher was invented, patented, commercialised and implemented, and how it has become one of the most important developments in security in the Information Age.

From a storyteller's point of view, there is a magnificent twist in the invention of public key cryptography. In 1997, the British government announced that researchers at the Government Communications Headquarters (GCHQ) had made the same breakthroughs as the American cryptographers, but ahead of them. However, the British research had been classified and the British researchers received no public credit for their work for a quarter of a century. The fact that the British inventors of public key cryptographer remained anonymous for so long contributes to a theme that runs throughout the book.

185

Throughout *The Code Book,* scientific explanations are surrounded by the stories behind the science. In my opinion, the background story of science is relevant to the science itself. Also, there are two main advantages to writing in the style of narrative non-fiction.

First, the story can create drama and tension which draws readers into the science. In other words, non-scientists may read narrative non-fiction whereas they might not read traditional science writing. The narrative structure may also give readers the momentum they require to get through some of the more technical sections. At the same time, readers who are familiar and content with traditional science writing do not seem to be perturbed by the addition of narrative detail.

The second advantage of narrative nonfiction is that adding stories to science writing can often mean the inclusion of history. I have found that a historical perspective is often helpful in introducing non-scientists to science, because the earliest stages of a scientific pursuit are generally easier to understand and provide a grounding for more complicated modern ideas. In *The Code Book,* the first chapter establishes the foundations of cryptography using various historical examples, whereas the final chapter is a description of quantum cryptography. Although it is complicated, my hope is that readers will feel confident enough to read about quantum cryptography because they have achieved a solid grounding while reading about the elementary historical ciphers.

I have been writing for only four years and have only two books to my name. In both cases, the narrative non-fiction approach was entirely natural. Before writing about science, I made science television programmes, and in order to appeal to a large audience I realised that I had to introduce narrative into my programmes. Hence, when I started writing, I translated my television style into my books.

Many other exponents of the narrative non-fiction approach to science writing seem to come from a similar background to my own. Authors such as Dava Sobel, Paul Hoffman (*The Man Who Loved Only Numbers*) and Sylvia Nasar (*A Beautiful Mind*) do not work in television, but they have had careers as journalists writing for newspapers and magazines, where storytelling is equally important.

For many authors and subjects, the narrative non-fiction style may not be appropriate. Greene may have been right to take a more traditional approach towards writing *The Elegant Universe*. String theory is an area of science without a long history, neither does it have rich characters around whom a story could easily be constructed, and what little story there is does not yet have an ending.

But in general, when authors are attempting to reach out to a general readership, I would encourage the use of story telling techniques where possible. Most popular science writers have the objective of explaining science to the layperson and raising awareness of scientific issues among the general public, and I believe that narrative non-fiction can help to achieve this. However, authors should always remember that science books are about explaining science, and therefore they should not forget to include explanations within narrative non-fiction. The danger is that the trend towards storytelling in science will go too far, and that some writers will be tempted to forget the science altogether.

186

Mathematics Takes Center Stage

Robert Osserman

It was just fifty years ago that mathematician Morris Kline, bothered by the general perception that mathematics was a subject divorced from the central concerns of society at large, took time out from his purely research activities to write the first of several books designed to demonstrate that precisely the opposite was true: mathematics is, and has been throughout history, absolutely central to a whole spectrum of human activities. His book *Mathematics in Western Culture* [6] appeared in 1953 and details not only the interweaving of mathematics with the sciences, but also with the arts, with philosophy, and with culture in general.

On the subject of Pierre de Fermat, whose highly original contributions to mathematics dazzled his contemporaries and earned him the title "Prince of Amateurs" by E. T. Bell [2], Kline writes "he lived an ordinary life as a lawyer and civil servant; at night, he indulged himself in mental sprees by creating and lavishly offering to the world million-dollar theorems."

Kline was obviously speaking metaphorically, at a time when the actual value of mathematical theorems was probably closer to five cents. He would have been incredulous had he been told that by the end of the century, there would be a number of mathematical theorems whose proofs would be *literally* worth a million dollars. Seven unsolved mathematical problems have been designated as "Millennium Prize Problems" by the recently founded Clay Mathematics Institute. At a widely publicized event held in Paris in the year 2000 to commemorate Hilbert's famous list of unsolved problems presented to the International Congress of Mathematicians in Paris in 1900, CMI issued their challenge and offer of a million dollars each for the solution of seven major problems, including the two oldest and perhaps most famous: the Riemann Hypothesis and the Poincaré Conjecture. In addition to those, an even older problem, the Goldbach Conjecture, was given a brief window as a literal million-dollar problem, when the publishers of Apostolos Doxiadis' book, *Uncle Petros and Goldbach's Conjecture*, offered a million-dollar prize for anyone who could find a solution within a limited period of time.

Perhaps nothing is more indicative of the radical change in the general view of mathematics than the surprising announcement that the annual „Jobs Rated Almanac" [7], a survey of public perception of job satisfaction, put statisticians and mathematicians in the first two out of 250 lines of work for "best working environment." The top ten choices for best overall job all involved some aspect of mathematics or computing. When Kline was writing, the average person would have been surprised to discover that mathematics was even a possible job category, rather than an academic subject from the distant past.

What accounts for the radical change not only in public awareness of mathematics but also in the shift from a rather negative stereotype of mathematicians to one that is far more favorable? There is no doubt that a major factor was the advent of computers. Oddly enough, the association between mathematics and computers in the mind of the public had little to do with the reality. For most people, mathematics *is* computing. The New York Times *Encyclopedic Almanac* for 1970, for example, has one page devoted to mathematics, a third of which is taken up with – of all things – a 25-by-25 multiplication table. It also includes the statement, "In recent years, computers have revolutionized mathematics." Nothing could have been further from the truth. In 1970 only a relative handful of research mathematicians had made any use whatever of a computer. They were much more likely to be part of the standard equipment of economists, psychologists, social scientists – anyone whose work involved dealing with large amounts of data, while they offered little to a mathematician devising new theories, defining new objects, trying to prove theorems or settle conjectures.

Much less known to the general public (and to the editors of the Times *Almanac*) was the real role of mathematicians vis-à-vis computers: that the basic ideas utilized in the construction of programmable computers were due to mathematicians like Turing and von Neumann. Furthermore, the important role of prototype computers in World War II was largely unknown, since much of it was long kept secret. But gradually that story emerged, along with the key role of mathematicians in the code-breaking efforts during the war. Again the name of Alan Turing appeared, in his heroic role in defeating the invincible-appearing German Enigma machine. The highly successful full-length biography, *Alan Turing: the Enigma*, by Andrew Hodges [5], appeared in 1983, and a large segment of the general public was treated to a view of mathematical activity that could only be described as significantly eye-opening.

It is a commonplace that art both reflects and influences society at large. During a period of transition, when broad changes are taking place, that dual role becomes self-reinforcing, creating a snowball effect. Something like that happened to the public perception of mathematics during the eighties and nineties, as an accelerating flood of books, plays, and movies depicted mathematicians, both real and fictional, and portrayed mathematical activities with varying degrees of accuracy.

Two events in the early nineties made headlines, and significantly increased public awareness of live mathematics. The first was the front-page news that Andrew Wiles announced his solution of the 350 year-old "world's most famous unsolved problem:" Fermat's Last Theorem. The second was the announcement that mathematician John Nash had been awarded a Nobel Prize for his early groundbreaking work on game theory. Both of those news items led to best-selling books: *Fermat's Enigma* [11] by Simon Singh (as well as several other books on the subject), and *A Beautiful Mind* [8] by Sylvia Nasar. In addition, there were two biographies of the amazingly prolific and idiosyncratic mathematician Paul Erdös.

But the greatest impact, and almost by definition the most dramatic, was that resulting from the ever more frequent depictions of mathematics on stage, in film and on television. Among the first was a play by Hugh Whitemore called "Breaking the Code" that was based on the Hodges biography of Turing. Derek Jacobi took the role of Turing both in the original production and in a later television version [16] that also features playwright and actor Harold Pinter.

When it comes to fictional mathematicians in the theater, the seminal work without question is the brilliant play "Arcadia" by Tom Stoppard [12]. The central character is the lively, charming, disarmingly brash young teenager Thomasina, who happens to be also a mathematical prodigy. Her tutor Septimus is well versed in mathematics and science, and much of their dialog, which takes place around 1810, reflects the latest developments in those subjects at that time. A parallel plot, set in the present day, involves a young population biologist whose role consists largely of interpreting the surviving 19th century notebooks, explaining them to the contemporary characters in the play, and relating them to his own use of modern mathematical methods. The typically intricate Stoppardian plot allows the interweaving of long bits of dialog devoted to a variety of sophisticated mathematical subjects, unlike anything else seen in an enormously popular play by a major playwright.

Stoppard did not put much significance in the fact that he had made his mathematical prodigy a young woman. But it is a curious fact that the majority of fictional mathematicians on stage and screen over the past two decades seems to have been women, in contrast to the ongoing concern about the scarcity of women in leading mathematics departments around the world. Some examples are:

"It's My Turn" (1980) in which Jill Clayburgh has the lead role as a research mathematician and professor at a major institution. She is portrayed in Hollywood fashion with an appealing personality, sex-appeal included, and she gets romantically involved with a major league baseball player (Michael Douglas). To add a touch of "reality" she has also to be a bit clumsy and absent-minded, but with a light touch.

"Presumed Innocent" (1991). Bonnie Bedelia plays a mathematics graduate student who has been working on her dissertation for ten years. Her husband Rusty (Harrison Ford) is the prime suspect in a murder case, but she is not above suspicion herself..

"Antonia's Line" (1995). Winner of the Academy Award for best foreign film of 1995. Five generations of strong-willed independent women. Antonia's daughter Danielle becomes a painter, and Danielle's daughter Thérèse is a mathematical prodigy, with every cliché in the book: a lightning calculator at an early age, interested in nothing at school except math and music, marries after getting pregnant, but has not much time or interest in her daughter Sarah, and only marginal interest in her husband. Sarah becomes a writer, and is the one who recounts the whole story. (It is true that one can interpret Thérèse's coldness in two ways – either as stereotypical math/scientist, or else as a consequence of having been brutally raped at a young age.)

By the time year 2000 rolled around, it may not be surprising in retrospect, but certainly seemed so a the time, when within a period of a few weeks three plays

189

opened in New York, all very different, but all having in common as a central character a young woman who is a kind of mathematical genius with a father who is also a prominent mathematician. One of them, "Hypatia", is based on the historical figure whose father Theon was a leading mathematician of the time, and who is the first well-known woman mathematician in Western history.

The other two plays are works of fiction. "The Five Hysterical Girls Theorem" by Rinne Groff is written in an absurdist style. Twelve of the eighteen characters are mathematicians, among them Moses Vaszonyi, an acknowledged mathematical genius. His daughter, not coincidentally named Hypatia, appears to have inherited his mathematical interest and talent. The dialog includes long stretches of mathematics, both real and imaginary.

The third play, "Proof" by David Auburn [1], was by far the most successful of the three. It moved on to Broadway in the fall of 2000, and received both the Pulitzer Prize for best dramatic work and the Tony Award for best new play. It has just four characters: Robert, a brilliant mathematician who was at the University of Chicago before suffering a nervous breakdown, Hal, one of his former students – now an instructor, and his two daughters Catherine and Claire. Claire is the only one who has left her father's mathematical sphere of influence. The central question of the play is whether Catherine has inherited her father's mathematical genius and/or his mental instability.

At the same time, male mathematicians and scientists were making their appearance with increasing frequency. More significantly, the way they were being portrayed was evolving rapidly. Reviewing three recent movies in which mathematicians played significant roles: "Presumed Innocent," "Sneakers" and "Jurassic Park", Constance Reid wrote in the Spring 1994 issue of Math Horizons, " The possibility of a star being a mathematician is very close to zero. The possibility of a realistic treatment of a mathematician is almost zero." On the second count she may be right, but along with everyone else, she must have been dumbfounded by what was just ahead for mathematicians in the cinema.

First in line was Jeff Bridges playing a math professor opposite English professor Barbara Streisand in the 1996 movie, "The Mirror has Two Faces". Immediately following was the 1997 movie "Good Will Hunting" in which Matt Damon plays a tough, brawling, bar-hopping denizen of South Boston who happens also to have a photographic memory and to be an untutored mathematical genius. The cult movie "Pi" from 1998, by director Darren Aronofsky, who went on to win a number of awards for his next film, "Requiem for a Dream," features another young mathematical genius who is subject to murderous migraine headaches, portrayed in chilling fashion.

In 2000 and 2001 Hollywood went all out in turning mathematical books into major motion pictures. The first was the film "Enigma", co-produced by Mick Jagger, and with a screenplay by Tom Stoppard. Based on the book of the same name by Robert Harris, its setting is the center of England's top-secret codebreaking efforts in World War II, Bletchley Park, and its chief protagonist is yet another mathematical genius, roughly modeled on Alan Turing. The second is based on the biography *A Beautiful Mind* [8]. It stars Russell Crowe of "Gladiator" fame as real-life mathematician and Nobel Prizewinner John Nash.

And so, mathematics has truly moved to center stage, in books and films as well as plays. The question that must be asked is: how accurate is the picture of mathematics and mathematicians that emerges out of all these depictions?

To start, let me note what seems to be a fundamental paradox that is of far broader scope than the particular case of mathematics. It is this: a writer, whether of biography or fiction, whether of books or plays, will naturally be drawn to a subject that is out of the ordinary, and more often than not, a story that is dramatic in some fashion. The reader, on the other hand, whose only exposure to a relatively esoteric subject is likely to be through these accounts, is almost guaranteed to form a picture that is correspondingly distorted. The question, then, is: if you had no other contact with mathematicians and if you diligently read and watched all these books, plays, and films over the past ten or twenty years, what picture would you form of mathematics and its practitioners?

The answer is glaringly obvious: mathematicians are either women or crazy – possibly both.

In the case of real-life mathematician John Nash, the portrayal is accurate. He was hospitalized for many years with a serious mental breakdown. In the New York Review of Books, under the heading "Varieties of Madness" [13], Joan Didion reviews the Nash Biography, *A Beautiful Mind*, together with the Unabomber manifesto. She points out that Time magazine refers to mathematician Theodore Kaczynski, the Unabomber, as a "mad genius," but that not everyone agrees to his insanity.

When it comes to fictional mathematicians, the record is impressive. The lead character, Max, in the movie "Pi," seems on the brink of a nervous breakdown, whether caused partly by his repeated excruciating migraines (or are they the result of his mental over-exertions?), by the confinement to his suffocating room with its enveloping patched- together computer system itself on the verge of permanent breakdown, or by his almost insane obsession with numbers.

Then there is mathematician Tom Jericho, the chief protagonist of the book and film, "Enigma" [4]. The book starts with Tom's mysterious return to Cambridge from Bletchley Park, sparking a storm of speculation, including "...he was a genius. He had had a nervous breakdown. ..." all of which we are told, was "precisely correct."

And then Uncle Petros, in the book of Doxiadis mentioned earlier [3], refusing to settle for a career in which he would be acknowledged as a brilliant and original mathematician of his time but without the crowning glory of solving one of the great open problems of mathematics such as Goldbach's Conjecture. He manages to more or less literally drive himself crazy in the attempt. His mathematically gifted nephew concludes that "with the Scylla of mediocrity on the one side and the Charybdis of insanity on the other, I decided to abandon ship."

Finally, a central theme of the play "Proof" [1] is the mental breakdown of the father and the potential of his daughter following suit.

The question, as we have said, is to what degree all these pairings of mathematical genius with insanity reflect reality.

First, the case of "Enigma" is easy to answer. Tom Jericho, the protagonist, is loosely based on Alan Turing, who seemed to get through all the boiler-room

191

pressures of Bletchley Park during the war without any hint of a breakdown. He is sometimes cited (in *Uncle Petros*, in particular) as an example of mental instability because he later took his own life. However, there is little evidence that the reason was connected to his mathematical genius, and much more cause to attribute it to the manner in which he was hounded because of his homosexuality, and his being forced to take hormones as a "cure."

As for "Proof," author David Auburn tells us (in Gussow[14]), "I think there is some connection between extreme mathematical ability and craziness. I don't think that math drives people crazy, but those with edgy or slightly irrational personalities are drawn to it." (A similar opinion is expressed in yet another piece of mathematically-oriented fiction [10]: "There is an ethereal quality in mathematics that has always attracted disturbed minds.") But *New York Times* reviewer Bruce Weber appears to have no doubt as to causation. He writes [15] about daughter Catherine: "she has witnessed firsthand the jumble that mathematics can make of a working brain."

There is no question that in the popular mind, too intense and sustained mental efforts such as those required for wrestling with mathematical problems can lead to mental breakdowns. That belief is both reflected and reinforced by passages in *Enigma, Proof,* and *Uncle Petros*. In fact, in the last two, the protagonists suffer relapses after trying to get back to their mathematical activities. The movie "Pi" is summarized on the cover of the video as "a brilliant mathematician on the brink of insanity as he searches for an elusive numerical code," and there is an older mathematician who warns his younger protégé against pushing too hard, and the danger of breaking down.

What is the reality of the connection between concentrated mathematical activity and mental breakdowns? To the best of my knowledge, there has not been any serious scientific study of the question. Anecdotally, in a lifetime devoted to mathematics, during which I have had contact with many hundreds of mathematicians, including most of the Fields Medalists, I can think of one or two who then suffered a mental breakdown. During the same period I knew a number of non-mathematicians who were forced to spend time in mental hospitals, and my visits there left me with the impression that mental illness strikes individuals from the entire spectrum of society, without regard to profession, race, or class. A recent book by Daniel Nettle [9] reports on evidence that there is a correlation between creativity, or "strong imagination" and various psychological disorders. Several studies seem to indicate that despite the stereotypical "mad scientist," comparisons across professions put scientists among the least subject to mental illness, with writers and playwrights at the top. Of course, the category "scientists" include all the experimentalists, whose success depends on sustained practical and organizational efforts which may be precluded by repeated bouts of mental disorder. Sylvia Nasar [8], citing Harvard psychiatrist John G. Gunderson, concludes "Men of scientific genius, however eccentric, rarely become truly insane – the strongest evidence for the potentially protective nature of creativity." None of the studies single out mathematicians or theoretical physicists, and some aspects of the studies themselves are open to question.

Whatever the case may be, the association between genius and insanity remains a strong one in the popular mind, and clearly a subject of continued fascination for writers.

Luckily, not all writers. Tom Stoppard has said [18] on the subject of his character Thomasina, the teen-age mathematical prodigy, "Rightly or wrongly, I mean accurately or inaccurately, I made her in every other respect a perfectly ordinary young woman. … The idea of genius in novels and art very often present us with a most unusual kind of human being. … I guess when one thinks of Gödel, for example, in life that may very well be also true, that you'd notice these people if you saw them on a train, and so forth. But I like to think that they are, perhaps, the egregious ones. And most very, very clever people – I find this an attractive idea – that if there's ten people messing around with a basketball in a court, one of them could be a genius, but you wouldn't know which by just looking at them and listening to them. And I like that aspect of that character."

Finally, the most improbable of all the recent representations of mathematicians in the theater is in the musical comedy, "Fermat's Last Tango," whose central character, Daniel Keane, is a fictional portrayal of Andrew Wiles. It retells the story of Wiles' initial announcement that he had finally found a proof of Fermat's Last Theorem, followed by the discovery that, in the words of one of the musical numbers, the "proof contains a great big hole." He then struggles to fill the gap, and – on the brink of giving up – finally succeeds. Despite the format of a musical comedy (and apart from the amusing fantasy of Fermat himself appearing to taunt and tease Daniel Keane) the story line is a remarkably accurate portrayal of what Wiles went through, with no compunction about avoiding mathematical language or formulas. Wiles himself is portrayed as a generally sympathetic character, with no suggestion that he was at the risk of a mental breakdown during the period when he was under enormous pressure, in full public view to either complete the proof or admit that he was unable to do so. And indeed, the same is true of the real-life Wiles, who, incidentally, is an even more sympathetic and appealing character then his stage portrayal. The authors of "Fermat's Last Tango," Joshua Rosenblum and Joanne Sydney Lessner had never met Wiles, and had only read his story in popular books.

The Clay Mathematics Institute, in addition to its offer of seven prizes of a million dollars each for the solution of seven mathematical problems, has done its bit to further mathematics in our culture by making a high-quality video recording of "Fermat's Last Tango" before the end of its original run in New York, which they have made available both on tape and DVD [17]. For a different view of mathematics and the theater, where mathematical genius is associated with song, dance, wit, and a little love interest, a viewer could do worse than to spend ninety minutes with this romp.

References

Books

[1] D. Auburn (2001) *Proof*, Faber and Faber, New York.
[2] E.T. Bell (1986) *Men of Mathematics,* Touchstone Edition, Simon and Schuster, New York, p. 56.
[3] A. Doxiadis (2000) *Uncle Petros and Goldbach's Conjecture*, Bloomsbury
[4] R. Harris (1995) *Enigma*, Random House, New York.
[5] A. Hodges (1983) *Alan Turing: The Enigma*, Simon & Schuster, New York.
[6] M. Kline (1953) *Mathematics in Western Culture*, Oxford University Press, Oxford.
[7] L. Krantz (2000) *Jobs Rated Almanac*, St. Martin's Griffin, reported in the San Francisco Chronicle, Sept. 1, 2000, pp. B1, B3.
[8] S. Nasar (1999) *A Beautiful Mind,* Touchstone Edition, Simon & Schuster, New York, p. 16.
[9] D. Nettle (2001) *Strong Imagination: Madness, Creativity and Numan Nature*, Oxford University Press, Oxford, pp.143-147.
[10] P. Schogt (2000) *The Wild Numbers*, Four Walls Eight Windows
[11] S. Singh (1997) *Fermat's Enigma*, Anchor/Doubleday, New York
[12] T. Stoppard (1993) *Arcadia*, Faber and Faber, London.

Articles

[13] J. Didion (1998) Varieties of Madness, *New York Review of Books*, April 23.
[14] M. Gussow (2000) With Math, a Playwright Explores a Family in Stress, *New York Times*, May 29, P. B3.
[15] B. Weber (2000) A Common Heart and an Uncommon Brain, *New York Times*, May 24, P. B1.

Videos

[16] Breaking the Code (1997), by H. Whitemore, Mobil Masterpiece Theatre.
[17] *Fermat's Last Tango* (2001), Clay Mathematics Institute.
[18] Mathematics in Arcadia: Tom Stoppard in Conversation with Robert Osserman, (1999) Mathematical Sciences Research Institute.

Additional book

[19] M. Emmer and M. Manaresi, eds. *Mathematics, Art, Technology and Cinema*, Springer-Verlag, Berlin 2003

Mathematics and Raymond Queneau

Michele Emmer

"Queneau never made a profession out of mathematics. He always did it for nothing; often with literature as his excuse" writes Jacques Roubaud at the start of his article *Les Mathématiques dans la Methode de Raymond Queneau* that appeared in the journal *Critique* number 359; Roubaud continues his analysis with a quote from the French mathematician, François Le Lionnais, a great friend of Queneau: "The idea of introducing undiluted mathematical notions into the creation of a novel or a poem came to us more or less just after we finished secondary school, during our time at university." [1] So, we're talking about the start of the nineteen twenties.

For Queneau, to be a mathematician meant, above all, to be a reader of mathematics: mathematical games (the famous pages in *Scientific American* by Martin Gardner); the history of mathematics (he was interested in the historical notes of the monumental work *Eléments de Mathématique* by Nicolas Bourbaki; famous mistakes (in *Bords*, [2] there's a chapter called *Conjectures fausses en théorie des nombres*); anecdotes; the "madmen" who take an interest in mathematics. In the book *Raymond Queneau: qui êtes-vous?* [3], Jacques Jouet publishes a previously unpublished work by Queneau entitled *Comprendre la follie*. Among the examples mentioned by Queneau are the classic cases of "proof" of squaring the circle, a problem that hundreds of "maths lovers" the world over continue to study. Even more amusing is the example cited in "Bords" [4] concerning Léopold Hugo, the nephew of Victor, who in 1877 published his "Hugodecimal theory, or scientific and definitive basis of the universal arithmelogistic that contains.... the panimaginary geometry of 1/m dimensions, arithmetic with 1/m digits, a presidential ecumenical decree regarding the Hugodefinitive basis of decimal numbering."

So, Queneau is: 1) a reader and 2) a "dilettante" of mathematics. These are the first two "propositions" that Roubaud "proves" in his essay on Queneau. Clearly the essay did not necessarily have to be written in this way, as if it were a mathematical paper, with "theorems" to be proved and "conjectures" that must be justified. The reasoning however is to be found in proposition 6, which states: he behaved in matters linguistic as if he were dealing with mathematics, bearing in mind that language is mathematifiable in as much as it is arithmetizable.

"Mathematics" however is too general a word; what kind of maths was Queneau interested in? The third proposition "proved" by Roubaud is the following: The favoured domain of Queneau is combinatorics.

But not exclusively. In 1948 Queneau published the essay *La place des mathématiques dans la classification des sciences* ('The role of mathematics in the classification of sciences) in the journal of *Cahiers du Sud* edited by his friend Le Lionnais, and dedicated to "Les grands courants de la pensée mathématique". The journal, which Le Lionnais had started to think about while in a concentration camp, gathered together the essays of famous mathematicians, of intellectuals and artists. An extended edition was republished in 1962 [5].

Though Le Lionnais edited the chapter *Les Mathématiques et la beauté* in which an essay by Le Courbusier also appeared, Queneau's contribution, rather tellingly, comes under the chapter *Les mathématiques et les sciences de la nature*. In the introduction to that chapter, Le Lionnais presents Queneau's essay, saying: "His curiosity is exercised in the most diverse fields of mathematics ... Taking mathematics as an end in itself, and not as a starting point for the other sciences, Queneau leaves open many great opportunities for revival." In his essay, Queneau writes: "The logical-mathematical system cannot simply be considered as the necessary and sufficient language of science, nor indeed as one of the sciences. To tell the truth, it is the Science par excellence ... Therefore, however one conceptualises the mathematification of the various scientific disciplines, one cannot doubt that the end aim of this transformation is the process of mathematification itself."

In 1948 Queneau joined the Société Mathématique de France. In July 1942 he started to write an essay entitled *Brouillon projet d'une atteinte à une science absolue de l'histoire* (Provisional project on the construction of an absolute science of history). In October of the same year, he abandoned the idea after writing only a few chapters. It was not until 1966 that the project, however incomplete, saw light with the title *Un histoire modèle* [6]. In the notebook Queneau "begins a contemplation of history on the basis of an emphasis of the relation between the numerical increase of a group of humans and the consequent fall in the amount of food available, ignoring any need to resort to work ([3], p. 96)." Chapter 65, with the typical title *Coefficients* reports: "In the mathematical theory of the lottery of life, a species is characterised by its coefficient of growth and by its coefficient of voracity. For a given group, other than its coefficient of cohesion and its coefficient of vitality, one can look at a coefficient of foresight and a coefficient of inventiveness."

"The mathematical study of two species, one the predator, the other the prey. We suppose that in an environment there live two species; if the first, the prey, were alone it would have a coefficient of growth that we assume to be constant and positive, call it e_1. If the second, which feeds itself wholly or mainly on individuals belonging to the first species, were alone it would have a coefficient of growth which we assume to be constant and negative, call it $-e_2$. When the two species coexist in a bounded environment, the first will develop far slower the more of the second species there are, and, in turn, the latter will develop far better the more of the former species there are. It is very simple to formulate the hypothesis that the coefficients of growth are of the form

$$e_1 - g_1 N_2$$
$$-e_2 + g_2 N_1$$

where g_1 and g_2 are positive constants, which leads to differential equations that give the variations in time of the two species

$$\frac{dN_1}{dt} = N_1(e_1 - g_1 N_2)$$

$$\frac{dN_2}{dt} = -N_2(e2 - g_2 N_1)$$

With suitable hypotheses, one obtains the periodic cycle law that says that the fluctuations of the two species are periodic. Furthermore, for certain numbers of the individuals, the state of the biological association is stationary and the equilibrium is stable."

The person writing this is a mathematician, an explicit fount of inspiration for Queneau and for his note-book: Vito Volterra. The volume is entitled *Leçons sur la Théorie Mathématique de la Lutte pour la Vie*, published in Paris in 1931 in the series *Cahiers Scientifiques*, edited by Gaston Julia (1893–1978) (the very same mathematician whose name was given to Julia sets in fractal geometry). It was a fundamental text for modern Biomathematics. In Queneau's note-book on chapter 20, entitled *The mathematical study of two species, one the eater, the other the eaten*: "It is shown that the fluctuations of the two species are periodic. For certain values of N and Q, (N the number of members in the group and Q the quantity of nutriment, or the number of the second group) the state is stable: there is no need to examine it here." In Queneau's note-book there appears not a single derivative sign, just a function sign, of the form Q(N)[1] .

In 1960 Queneau, together with Le Lionnais, founded the OULIPO, or Ouvroir de Littérature Potentielle, which was originally called Selitex, Séminaire de littérature expérimentale. "Queneau's far greater than amateur interest in mathematics", writes Jouet "was the principal driving force of the OULIPO." In the OULIPO's first manifesto, the "Oulipian" guiding concept of constraint is introduced: "Every literary work constructed, starting from an inspiration must, for better or for worse, follow a series of constraints and processes that each contain the other, like Russian dolls." Without going into too much detail, (for which see Roubaud's article), a good "Oulipian" constraint must be simple, the choice of constraints must not be accidental; a constraint is a type of axiom for the text. Roubaud's proposition 15 goes as follows: "Writing by 'Oulipian' constraints is the literary equivalent of writing a formalised mathematical text according to the axiomatic method".

As Calvino notes [7] "I would like to insist that, for Perec, constructing a story based on fixed rules, 'contraintes' (the same word used by the OULIPO), did not suffocate narrative freedom, but stimulated it. ... Queneau ... wrote: "Another extremely untrue idea that is also currently in vogue is the equivalence that is cre-

---

[1] *A comment in a daily paper at the Italian re-print of the little volume spoke of "extravagant numerical calculations". There were no calculations in Queneau's book; just some observations that are extended to human history, some calculations made by Volterra, calculations that one could scarcely call "extravagant".*

ated between inspiration, exploration of the subconscious, and liberation; among them autonomy and liberty. Now, this inspiration that involves the blind following of every impulse is in reality a slavery. The classicist who writes his tragedy observing a certain number of rules is more free than the poet who writes whatever passes through his head and is a slave to other rules of which he is unaware." (from *Bâtons, chiffres et lettres*, 1950)". The Oulipo's method of constraints reminds us immediately of another method which was very popular between 1940 and 1960, in the mathematical community: the axiomatic method, in particular the work of Nicolas Bourbaki.

"The axiomatic method", writes Bourbaki "to be clear, is none other than the art of compiling some texts whose formulations are easily understood. This is not a new invention, but its systematic application as a research tool is one of the original works of contemporary mathematics. In reality, when one is writing or reading a formalised text, it doesn't matter if we assign determined meanings to the letters and signs of the text or no meaning at all; all that matters is the correct abidance by the rules of the syntaxes." (from *Poésie des ensembles*). In *Storia della matematica* (The History of Mathematics) [8], Boyer calls him the "Polycephalic Mathematician". In fact, there exists no mathematician by this name. Bourbaki is the Greek name of a non-existent Frenchman, that appears on the front page of several tens of volumes that form a series entitled *Eléments de mathématique*, and which aim to set out all that is important in mathematics. Nancy (where there is a statue of general C. D. Bourbaki) is said to be his birth place; his university that of Nancago, a play on the words University of Chicago, where several of the mathematicians of Bourbaki worked. Among the group of mathematicians were H. Cartan, C. Chevalley, J. Dieudonné, C. Ehresmann and A. Weil. The first volume by Bourbaki was published in 1939. To date more that thirty such volumes have been published. Besides the collective article by Bourbaki, *L'achitecture des Mathématiques*, in Le Lionnais' volume there are articles by Dieudonné and Weil. In his two discussions, Dieudonné spoke of the mathematician David Hilbert (1862–1943) and of the modern methods of axiomitization at the heart of mathematics.

In 1899 Hilbert had published a book called *Grundlagen der Geometrie* (Fundamentals of Geometry) which had become famous. By merit of this work, Hilbert became the principal exponent of the axiomatic school of mathematical thought. Boyer writes: "The order of the geometry developed by Hilbert. ... put into perspective the fact that those terms not defined by geometry must be assumed without attributing other properties to them beyond those indicated in the axioms. One needed to abandon the empirical-intuitive level of old geometric ideas, and one had to conceptualise points, lines and planes simply as elements of certain given sets. ... Analogously, undefined relationships had to be considered as simple abstractions that indicated nothing but a correspondence or a representation."

Hilbert's text couldn't fail to fascinate Queneau, who in 1976, shortly before his death on the 25th of October of the same year, published "Les fondaments de la littérature d'après David Hilbert". The method used by Queneau was described by him as follows: "Taking inspiration from this illustrious example, I set out an axiomatic approach to literature by substituting the words 'points', 'lines', 'planes' in Hilbert's propositions with the words 'words', 'sentences' and 'paragraphs' respectively." Here is an example of Queneau's axioms: "Given a sentence, consider a word not belonging to that sentence; in the paragraph determined by the given sentence and word, there exists at most one sentence that contains the given word and that has no word in common with the given sentence" which is some sort of equivalent of Euclid's parallel postulate.

One of Queneau's greatest interests was for combinatorics associated with, in particular, whole numbers, a type of "arithmania" in which he displays a completely Hellenistic faith in the birth of formal harmony via numbers. The 1961 work *Cent mille milliards de poèmes* (One Hundred Million Million Poems) typifies this point of view. The principle is the following: one writes ten sonnets with the same rhymes; the grammatical structure is such that every verse of every "base" sonnet is interchangeable with every other that is in the same position. So, for every verse of a new sonnet one composes, one has ten independent possible choices. There are 14 verses; one therefore has, potentially, 10^{14} choices, or one hundred million million sonnets, which, Queneau notes, would take one hundred million years to read.

His interest in numbers is also supported by a short film called *Arithmétique*, a film not cited in his biography of 1988. The film was probably made sometime between 1955 and 1960. The subject is arithmetic and the properties of whole numbers. Adopting a deadly serious expression, at times as if in a police thriller, and interspersed with blasts of trumpet and surreal images, Queneau states absolutely correct and true properties, juxtaposed with paradoxical observations and witticisms, of whole numbers.

But did such a scholar, so attracted by mathematics, a friend of mathematicians, a member of the French society of mathematicians, ever have the ambition at least once to write a mathematical work, with original results, in a mathematical journal, aimed at the mathematical community? A sort of test of his own ability to be a mathematician, if he so wished ... The answer is obviously yes, and given the circumstances it could be nothing but a work on the theory of numbers, in particular, the integers.

In 1968 a certain note of his was accepted by the Paris Academy of Sciences. The brief note, which was presented, as was customary, by a mathematician member of the Academy, in this case André Lichnerowicz, was discussed in the sitting of the 29th April and published in the Comptes Rendus (proceedings) of the Academy on the 6th May 1968 [9]. The subject of the note was s-additive sequences. At the start, Queneau recalls the definition of an s-additive sequence: it is a strictly increasing sequence of positive, whole numbers, in which every number is the sum in s different ways, of two different numbers that belong to the sequence. For example: starting from a base of 1, 2, 3, 4 the next number will be $5 = 3 + 2$ or $5 = 4 + 1$. Then 6. Not 7, as $7 = 5 + 2 = 6 + 1 = 4 + 3$, and can thus be written in three different ways and not just two. The results of Queneau's note, as with all mathematical works in the world, were examined in Mathematical Reviews in 1969. The final work in definitive form was then published in the *Journal of Combinatorial Theory*, [10], edited by another famous mathematician, Gian-Carlo Rota. The work's title is *Sur les Suites s-additives*.

As well as giving results, the work makes some conjectures that the author is not able to prove. The observation Queneau makes on page 64 of his article is telling: "... one is happy to recover the Fibonacci series (1180–1250)", a series of several centuries' fame, related to the "golden section". Roubaud comments: "Queneau's interests in the combinatorics of whole numbers are a conscious decision, interests that make him part of a tradition as ancient as western mathematics." For once the scholar wanted to see how it felt to "prove" a mathematical result which, being free of errors, was indisputable. Who knows, perhaps Queneau considered this the best example of literature with "Oulipian" constraints?

References

[1] F. Le Lionnais, *Raymond Queneau et l'amalgame des mathématiques et de la littérature.* La Nouvelle Revue Fran‚caise, 290, p.76, Feb. (1977)

[2] Queneau, R.: *Bords.* Hermann, Paris (1963)

[3] Jouet, J.: *Raymond Queneau: qui êtes-vous ?* Ed. La Manifacture, Lyon (1988)

[4] In [3], p. 82

[5] Le Lionnais, F.: *Les grands courants de la pensée mathématique.* Ed. A. Blanchard, Paris (1962)

[6] Queneau, R.: *Un histoire modèle (1966),* Italian edition: *Una storia modello,* Einaudi (1988)

[7] Calvino, I.: *Lezioni americane.* Garzanti, p. 119. (1988)

[8] Boyer, C.: *Storia della matematica.* Italian edition: ISEDI, Milano (1976)

[9] Queneau, R.: *Sur les Suites s-additives.* C. R. Acad. Sc. Paris, (A) 266, 957–958, May (1966)

[10] Queneau, R.: *On the s-additives series.* Journal of Combinatorial Theory, (A) 12, 31–71 (1972)

Authors

Thomas F. Banchoff

Department of Mathematics, Brown University
Providence, RI 02912
thomas banchoff@brown.edu

Davide P. Cervone

Department of Mathematics, Union College
Schenectady, NY 12308
dpvc@union.edu

Colin Clifford

School of Psychology, Sydney University, Sydney
Australia

Apostolos Doxiadis

Athens, Greece

Michele Emmer

Dipartimento Matematica "G. Castelnuevo"
Università di Roma "La Sapienza"
Piazzale A. Moro, 00185 Roma
emmer @mat.uniroma1.it

Anatoly T. Fomenko

Department of Mechanics and Mathematics
Chair of Differential Geometry and Application
Moscow State University, 119899 Moscow, Russia

George K. Francis

Department of Mathematics
University of Illinois
gfrancis@uiuc.edu
http://www.math.uiuc.edu/#gfrancis/

Herbert W. Franke

Austraße 12, Puppling
82544 Egling
franke@zi.biologie.uni-muenchen.de

Dietmar Guderian

University of Education Freiburg
Pädagogische Hochschule, Kunzenweg 21
D – 79117 Freiburg im Breisgau

Alexandr O. Ivanov   *Department of Mechanics and Mathematics*
Moscow State University, 119899 Moscow
Russiaaoiva@mech.math.msu.su or
tuz@mech.math.msu.su.

Michael Joswig   *Technische Universität Berlin*
Institut für Mathematik
Straße des 17. Juni 136, 10623 Berlin
joswig@math.tu-berlin.de

Judith Flagg Moran   *Mathematics Center, Trinity College*
Hartford, CT 06106-3100 USA
judith.moran@mail.trincoll.edu

Ben Newell   *Department of Psychology, University College*
London, London, UK

Robert Osserman   *Mathematical Sciences Research Institute*
Berkeley, CA 94720-5070
ro@msri.org

Achille Perilli   *Largo Arenula 34, 00196 Rome, Italy*

Konrad Polthier   *Fachbereich Mathematik,*
Technische Universität, Berlin
polthier@math.tu-berlin.de

Michea Simona   *Accademia di architettura, Università della Svizzera*
italiana, Largo Bernasconi 2, CH-6850 Mendrisio
msimona@arch.unisi.ch

John Sims   *Coordinator of Mathematics*
Ringling School of Art and Design
2700 North Tamiami Trail
Sarasota, Florida 34234, USA
jsims@ringling.edu

Simon Singh   *P.O.Box 23064, London, W11 3DQ, England, UK*
simonsingh@visto.com

Branka Spehar   *School of Psychology, University of New South*
Wales, Sydney, Australia

Richard Taylor   *Physics Department, University of Oregon*
Eugene 97403-1274, USA
rpt@darkwing.uoregon.edu

Alexey A. Tuzhilin

Department of Mechanics and Mathematics
Moscow State University, 119899 Moscow, Russia
aoiva@mech.math.msu.su or
tuz@mech.math.msu.su.

Kim Williams

Nexus Network Journal
Via Mazzini, 7 – 50054 Fucecchio (Firenze) Italy
k.williams@leonet.it